The New
Enzyme-Catalyst Diet

Amazing Way to Quick,

Permanent Weight Loss

The New
Enzyme-Catalyst Diet

Amazing Way to Quick,
Permanent Weight Loss

Carlson Wade

Foreword by
William S. Keezer, M.D.

PARKER PUBLISHING COMPANY, INC. West Nyack, N.Y.

Library of Congress Cataloging in Publication Data

Wade, Carlson.
   The new enzyme-catalyst diet.

   Includes index.
   1. Reducing diets. 2. Enzymes--Therapeutic use.
3. Plants, Edible. 4. Food, Raw. I. Title.
RM222.2.W2   613.2'5   76-3557
ISBN 0-13-613349-5

Dedication

To Your Brand New Youthfully Slim Body

Other Books By The Author

Helping Your Health With Enzymes
Magic Minerals: Key to Better Health
The Natural Way to Health Through Controlled Fasting
Carlson Wade's Gourmet Health Foods Cookbook
Natural and Folk Remedies
The Natural Laws of Healthful Living
Health Tonics, Elixirs and Potions for the Look and Feel of Youth
Natural Hormones: The Secret of Youthful Health
Health Secrets from the Orient
Magic Enzymes: Key to Youth and Health
The Miracle of Organic Vitamins for Better Health
Miracle Protein: Secret of Natural Cell-Tissue Rejuvenation
All-Natural Pain Relievers

FOREWORD
By a Doctor of Medicine

For the first time ... an incredibly *quick* and *foolproof* way to take off all the overweight you want—permanently! Here is a major new discovery—a weight-loss revolution. It actually melts the fat right out of your body—almost overnight! It is the proven alternative to most other diets ever used.

Carlson Wade, an internationally known medical researcher and author, has written an exciting new book about a little-known major breakthrough in the problem of losing weight and keeping it off.

It is called the Enzyme-Catalyst Diet. It is a miracle in the constant search for a safe, healthy and satisfying reducing plan. Now, with Carlson Wade's dynamic new Enzyme-Catalyst Diet plan, you go right on eating and enjoying most of your favorite foods and still lose overweight. Yes, even if other reducing methods have failed, you can now lose as much overweight as you want—and keep it lost—with this new Enzyme-Catalyst Diet plan.

This book describes the amazing discovery of the search for a successful weight losing method. Carlson Wade shows how with this Enzyme-Catalyst Diet, you no longer need to suffer hunger pains. No complicated formulas to follow. No special foods to prepare. No counting calories or carbohydrates with every meal. More important, *No drugs, no pills, no exercises!*

Instead, you are encouraged to fill yourself up with many delicious, luscious, lip-smacking good foods. That's right. You can enjoy many favorite foods that contain little-known, newly-discovered weight-melting factors—enzymes! When you eat these

7

tasty good foods, you automatically create a metabolic-catalyst action that will actually "wash out" 10-20-30-50-70-100 pounds of weight from your body cells.

Your reward? You will become permanently slim on this plan, without any extra effort on your part . . . other than eating the delicious foods described in the book.

Sound unbelievable? But it's true. In reported case history after case history, you will read about overweight folks who have shed "mountainous" pounds of ugly fat under this Enzyme-Catalyst Diet plan. It has succeeded where nearly all other strenuous, tortuous and unhealthy diets have failed.

Carlson Wade's Enzyme-Catalyst Diet plan is the only all-natural program that keeps you in good health, satisfies your appetite, while it takes pound after pound of overweight out of your body. It makes dieting a delicious pleasure!

Most welcome news of all, the Enzyme-Catalyst Diet plan does away with the single greatest cause of weight-loss failure—will power. That's right! With the ECD Plan, you no longer have to fight the urge to eat. You will discover a natural freedom from a compulsive eating urge. Enjoy the described enzyme-catalyst foods and you will feel delightfully satisfied after your meals. Moments after you have finished eating, your metabolic system will start catalyzing excess weight and melt overweight pounds right out of your body.

Carlson Wade has written a no-nonsense book for folks who want a quick and permanent weight-loss guide. He tells you (1) How to lose weight fast, (2) how to keep it off permanently, (3) how to slim down without any drastic starvation-punishment diets or dangerous drugs and pills.

Even the most stubborn cases of overweight respond to this miracle diet plan. And that's just the beginning. Carlson Wade's Enzyme-Catalyst Diet plan gives you these miracle weight-losing programs:

- How to use special tonics to create a natural appetite control.
- How to use everyday foods in a unique way to help decontrol your compulsive eating urges.
- How 25 everyday foods can act as all-natural, organic reducing pills.
- How to use enzymes to protect against flabby skin while slimming down.
- How to enjoy good foods and keep slim while dining out.

- How to use enzyme foods to soothe those "runaway glands" that lead to overweight.
- How to follow a Two Week ECD plan to become permanently slim.
- How to use 50 different ECD slimming secrets for daily weight loss.
- How to enjoy variations with 15 different ECD programs for fast, permanent weight loss.

A special bonus is the Enzyme-Catalyst Diet Cookbook. Gourmet recipes. Easy to prepare. Deliciously enjoyable. Yet they trigger off a catalyst action in your metabolic system that creates miracle weight loss in a short time . . . while you enjoy favorite good foods.

Carlson Wade is a highly respected writer known to millions the world over through his many articles, books, and as editor of the internationally known *Parker Natural Health Bulletin*. Now, with this miracle discovery of the Enzyme-Catalyst Diet, he offers long-awaited hope for permanent weight loss to millions of overweight. It is the latest and most effective natural way to lose weight . . . and keep it off! It is highly recommended!

William S. Keezer, M.D.

What This Book
Will Do for You

This book is written for the 60 million overweight Americans who are faced with the never-ending problem of taking off unhealthy fat—and keeping it off!

If you are one of these overweights, you want to lose unsightly pounds. But you do not want to sacrifice your favorite foods. Instead, you want to enjoy many delicious meals daily, and still slim down. You want to continue eating your favorite stews, succulent steaks and chops, steaming casseroles, favorite sweet desserts ... while you get rid of inches, bulges, "spare tires" from all parts of your body.

Neither do you want to suffer through starvation regimens, drugs and pills, mechanical "reducers," complicated and monotonous exercises. But you still want to lose weight!

Isn't there any diet for you?

Yes! This book is a step-by-step guide to the *lifetime stay-slim plan* you've always wanted but could not find. There are *no* special foods. *No* complicated exercises. *No* pills. *No* drugs. But most important, *no will power!*

That's right. This book is different from all others because it lets you forget about your will power ... shows you how to lose weight with almost NO effort on your part.

This book will show you how to continue eating most of your favorite foods on a unique Enzyme-Catalyst Diet that sheds pounds on a delicious *no will power* program. Could you ask for anything better?

11

This book offers hope for the 60 million (and increasing) Americans who are victims of poor will power and runaway appetites. As one of them, you have probably suffered through the hunger pangs of other diets, backbreaking calisthenics, dizzying-sickening drug reactions, monotonous group therapy clubs ... only to continue tipping the overweight scales. You have been searching for a diet program that lets you enjoy good foods ... while you slim down.

This book offers you exactly that type of diet program. It shows you how certain everyday foods can create a metabolic-catalyst reaction within your body (with no effort on your part) so that thousands of enzymes can actually melt the fat right out of your billions of body cells. This keeps you "forever slim" while you enjoy favorite gourmet foods.

This book shows you how to keep on eating fats, carbohydrates, sweets, snacks and so-called forbidden "goodies" on an Enzyme-Catalyst program that helps you lose weight with every juicy good bite and swallow.

The ECD Program is more effective and successful than most so-called diets because:

- It makes you feel energetic, youthful, instead of nervous or tense as with other punishing diets.
- It lets you eat most of your favorite foods, without the need for changing your eating habits.
- It frees you from the problem of will power, a common cause of most other diet failures.
- It lets you eat favorite foods at home, at parties, in restaurants, without the embarrassing restrictions other diets impose upon you.
- It takes weight off permanently and helps you look and feel "forever slim." (Most diets have a "yo-yo" effect of putting weight back on as quickly as it was taken off.)

This book shows you that dieting can be delicious! You'll taste the joys of eating your way to a healthy slim shape. The Enzyme-Catalyst Diet Program makes reducing a whole lot of fun!

This book will help you reduce even if you complain your fat is in your glands, that you inherited your overweight, that you have an urge to eat when feeling "blue," or that you are just the "fat type." These are excuses that other diet programs tend to overlook. Not so with this book which explains many overweight causes and shows you how you can easily correct them ... with almost no effort on your part ... while you keep on eating!

With the Enzyme-Catalyst Diet Program, reducing is a delicious adventure that will help you look better, live longer with your lifetime stay-slim shape.

Added Bonus Feature: As a special feature, this book offers a **Diet-O-Matic Index.** This is *not* found in any other reducing book. If you have a specific difficult weight problem, look it up in the index to locate the program and solution. This **Diet-O-Matic Index** gives you "instant" advice and programs on many weight conditions. All at-a-glance.

So now, take the first step. Go to the first chapter. You stand on the threshold of a new youthful "forever slim" body ...

Carlson Wade

Contents

1

How to Set Your Weight Goal with the ECD Slimming Plan

The moment you opened up this book, you took your first step toward losing excess weight and slimming down to better health and more youthful looks. You showed that you want to melt away those extra pounds. This is important. You must be motivated to lose weight. Once you take this initiative, you can then cooperate with Nature to help you melt away excess pounds and use the Enzyme-Catalyst Diet to "eat and become slim." *Let Nature melt away your pounds.* This is the fast, permanent weight loss program that will help you look younger, feel better, enjoy longer years of a youthful lifespan. The Enzyme-Catalyst Diet (or ECD, as it is called) helps you enjoy everyday, *healthful* foods while it performs an internal fat- and calorie-dissolving action that helps "melt" and "wash" away excess pounds. *It is the all-Natural way to become slim and stay slim for the rest of your healthful life.*

Get Ready To Lose Excess Weight. The moment you started reading this book, you made a commitment with yourself. You accepted your responsibility to lose excess weight. Look upon the ECD Program as a tasty and exciting adventure. The ECD Program offers you an endless variety of many delicious and interesting foods that satisfy your appetite, nourish you and *slim you down*. While the ECD Program calls for moderation in some foods and substitutions of others, it is unique and successful in one basic way: *it shows you how certain everyday foods, eaten singly or in combination, can act as reducing pills!* You can actually "feast" your way to a slim-trim and healthful body. Get ready to enjoy new satisfaction, new pride and well-deserved pleasure when you see bulges around your middle (and elsewhere) start to disappear as your weight melts down. All this and much, much more with the ECD Program. Here's what it can do for you:

HOW THE ENZYME-CATALYST DIET LETS YOU EAT YOUR WAY TO A SLIM-TRIM SHAPE

Fresh raw fruit, vegetables, grains, seeds, nuts, and their juices are powerhouses of delicate life-like substances called *enzymes*. When you eat a variety of enzyme-containing foods, in amounts, schedules and combinations to be described in this book, you give your metabolism a treasure of these substances.

Alerts Digestion, Metabolism, Assimilation. When you eat enzyme foods, these substances speedily boost your powers of digestion, adjust your metabolism with the precision of a built-in body "clock" and then help to "burn up" or create "assimilation" of the fats, carbohydrates, calories and other substances. Without sufficient enzymes, these compounds become deposited in your body and weight builds up. Enzymes act as spark plugs within your digestive system, providing your body with needed energy, giving your digestive system's "furnace" the "fuel" it needs to "burn away" unwanted pounds.

Creates "Body Combustion." Enzymes act as catalysts. That is, they are able to stimulate an internal reaction without themselves being transformed and destroyed in the process. This enzyme-catalyst action creates "body combustion" wherein fats, carbohydrates, calories, etc., are "burned" to provide fuel for your body. When you have an adequate and daily amount of enzymes in your system, they act as a natural "stop gap" to guard against weight build-up. A deficiency of enzymes means a slowing down of your "body combustion." Accumulated fats, carbohydrates, calories start to build up. They cannot be adequately catalyzed or metabolically "burned" and the weight problem begins to mount. The key to better health, the goal for a fast, permanent weight loss begins with enzyme nutrition!

Enjoy Good Foods With Enzyme Content. You can enjoy adequately satisfying meats, dairy foods, stews, goulash dishes, casseroles, French toast, desserts, snacks, nibbles, delicious goodies, eggs, and almost any wholesome meals, provided you follow this basic ECD Program rule: *each and every meal should contain an enzyme containing food.* As soon as you follow this basic rule of thumb, you can start looking forward to the enjoyment of good foods . . . as you slim down. This book will show you *how you can feast high while you slim down,* with the ECD Program.

HOW TO SET YOUR WEIGHT GOAL

Decide how much weight you need to lose. To find out *if* you are overweight, here are two methods:

1. *Consult The Chart.* The chart that follows is a general guide to best (that is, healthy) weights for men and women at age 25 and over. Since people come in a variety of sizes and shapes, large-boned or small-bones, firm muscled or flaccid, short and stocky or tall and lanky, and many variations in between, no one weight is right for everyone of the same height and sex. The weight that is best for you depends upon your individual frame size and muscular development. It is the weight at which you look and feel your best. Some people look and feel better when they weigh somewhat more than their statistically-desirable weight, if that extra weight is largely firm muscle, not fat.

2. *The Pinch Test.* To determine if too much of your weight is just fat, pinch the back of your upper arm. If you can pinch a thickness of one inch or more, chances are you are carrying excess fat—and excess weight. (This test is most meaningful for persons under 50 years of age.) If you are 20 percent above your desirable weight, your physician will probably recommend that you start losing weight. With the ECD Program that follows, this will happen easily, quickly and permanently.

Make Adjustments On Chart. Please note, too, that the weights given are for persons wearing indoor clothing and shoes. The heights allow for shoes with one-inch heels for men, and shoes with two-inch heels for women. If you usually weigh yourself on your bathroom scale without shoes (and perhaps without any clothing) these factors must be taken into account, and adjustments made before you find your desirable weight. For nude weights, women should subtract 2 to 4 pounds, men should subtract 5 to 7 pounds.

Weight in Pounds According to Frame (In Indoor Clothing)

	HEIGHT (with shoes on) 1-inch heels Feet Inches		SMALL FRAME	MEDIUM FRAME	LARGE FRAME
Men *of Ages 25* *and Over*	5	2	112-120	118-129	126-141
	5	3	115-123	121-133	129-144
	5	4	118-126	124-136	132-148
	5	5	121-129	127-139	135-152
	5	6	124-133	130-143	138-156
	5	7	128-137	134-147	142-161
	5	8	132-141	138-152	147-166
	5	9	136-145	142-156	151-170
	5	10	140-150	146-160	155-174
	5	11	144-154	150-165	159-179
	6	0	148-158	154-170	164-184
	6	1	152-162	158-175	168-189
	6	2	156-167	162-180	173-194
	6	3	160-171	167-185	178-199
	6	4	164-175	172-190	182-204

	HEIGHT (with shoes on) 2-inch heels Feet Inches		SMALL FRAME	MEDIUM FRAME	LARGE FRAME
Women *of Ages 25* *and Over*	4	10	92- 98	96-107	104-119
	4	11	94-101	98-110	106-122
	5	0	96-104	101-113	109-125
	5	1	99-107	104-116	112-128
	5	2	102-110	107-119	115-131
	5	3	105-113	110-122	118-134
	5	4	108-116	113-126	121-138
	5	5	111-119	116-130	125-142
	5	6	114-123	120-135	129-146
	5	7	118-127	124-139	133-150
	5	8	122-131	128-143	137-154
	5	9	126-135	132-147	141-158
	5	10	130-140	136-151	145-163
	5	11	134-144	140-155	149-168
	6	0	138-148	144-159	153-173

For girls between 18 and 25, subtract 1 pound for each year under 25.

HOW MUCH WEIGHT DO YOU WANT TO LOSE?

Based on the desirable weight tables, your own estimate of your determination to lose and your doctor's advice, setting your weight goal is the next step.

Let's say that you and your doctor have decided that you should lose a total of 35 pounds. You might try to lose half this amount over a period of two or three months. During this time, you will attempt to shed *about two pounds a week*—sometimes more, sometimes less. The balance will be more gradual. The ECD Program makes it easy for you to lose weight while you continue eating most of your favorite foods. Set yourself a goal and then use the ECD Program to accomplish that slim-trim figure you've always wanted.

KEEP A RECORD OF YOUR WEIGHT CHANGES

You will be able to see the rewards of the ECD Program if you keep a record of your weight changes. Copy the following chart and then mark down your weight changes from week to week. You will then be able to determine whether you should include more enzyme foods in your programs or make other adjustments in portions of foods, as described throughout this book. It is the easy way to chart your own progress right in the privacy of your home.

HOW THE ECD PROGRAM CREATES A SLIM BODY BY REDUCING YOUR "FAT" CELLS

Your body is composed of billions upon billions of cells and tissues. Each cell is a gelatinous substance. Each cell has a dense kernel, the *nucleus,* which directs the activity of the rest of the cell. Within the nucleus of the cell is an ingredient called the *mitochondria.* This ingredient influences combustion and metabolism. If the *mitochondria* is adequately nourished with a catalyst such as the enzyme, it can then promote better combustion so that fats, carbohydrates and calories can be better metabolized. A deficiency of enzymes tends to weaken the *mitochondria* so that it cannot create the "fat burning" process that helps keep you slimmer. Pounds can then build up because eaten fats, carbohydrates and calories *accumulate,* instead of being metabolized. The key to a slim body calls for an Enzyme-Catalyst Diet program that acts to *ignite* the *mitochondria* so that it can create the combustion needed to metabolize weight causing fats, carbohydrates, calories.

MY WEIGHT

**A bathroom scale is a must to keep track of your weight.
Weigh at the same time each day or week.**

Month	First Week	Second Week	Third Week	Fourth Week	Pounds Lost
March					
April					
May					
June					
July					
August					
September					
October					
November					
December					
January					
February					

Enzymes Create Cell Washing Action. Basically, overweight can be traced to your "fat cells." The amount of fat cells you have remain constant from puberty. Once you become an adult, the *number* of fat cells that you have, remain in your body and cannot be changed.

The problem is that if you are enzyme-deficient, these fat cells keep increasing in *size.* The nucleus of the cell accumulates more and more padding. As your fat cells swell up, so do you. Your fat cells need a "washing" whereby the excess fat will be oxidized, or "burned off" so that you can lose weight. Enzymes perform this catalyst action. They actually "wash" away the fat from your cells, thereby creating a natural and healthful weight loss.

Use Enzyme Foods In Your Weight Goal. When you have determined how much weight you need to lose, and begin to use the Enzyme-Catalyst Diet, it is important to eat *raw fruits, raw vegetables, raw seeds, raw plant juices,* with each and every one of your daily meals. These foods are prime sources of enzymes which then are used by your billions of body cells to help alert the *mitochondria* to perform internal combustion; this action helps to melt the accumulated fat in your cells, and wash them right out of your body. The ECD Program goes to the *cause* of your overweight; namely, fat-enlarged cells. By slimming down your fat cells, your body is rewarded with a healthful and natural slim-trim figure.

Basic Slimming Rule: Begin each meal with a raw fruit or vegetable salad. End each meal with a raw fruit salad. Enzymes in the raw fruits and vegetables are the vital spark plugs needed by your body to promote *cellular metabolism* to slim down your fat cells . . . and slim you down, too!

HOW THE ENZYME-CATALYST DIET PROGRAM CAN HELP SAVE YOUR LIFE

When you lose weight under the ECD Program, you do more than just look and feel healthfully trim. You extend your lifeline. You protect yourself against illness. The ECD Program can actually save your life. Here is what the ECD Program does for you, by helping you shed pounds:

1. *Protects Your Heart.* The risk of a serious heart attack is much greater for an overweight than someone who is slim.

2. *Soothes Your Blood Pressure.* Also known as hypertension, it is about twice as common in overweight folks than those of healthful weight. When you slim down under the ECD Program, your blood pressure is pampered and soothed.

3. *Hardening Of The Arteries.* Also known as athersclerosis, this condition is characterized by the formation of fatty deposits in the inner layer of the arteries. It is more common in overweight folks. When the ECD Program slims you down, the fat cells are "washed" so that the deposits are removed and there is protection against this ailment.

4. *Guards Against Diabetes.* A metabolic disorder of carbohydrate metabolism, it is much more common in overweight folks than those

of healthful weight. Enzymes help metabolize carbohydrates so that body cells are healthful and there is less predisposition to sugar buildup and the risk of diabetes.

5. *Helps Shield You From Osteoarthritis.* This ailment affects the weight-bearing joints of the body (lower spine, hips, knees) and is more common in overweight people because of the extra strain put on the joints. An ECD Program that takes off excess weight helps relieve the pressure on your joints. The symptoms are less severe. The development of the ailment is checked.

6. *Promotes Better Health.* Overweight also contributes to gall bladder disease, malfunctioning of the lungs. *It can even threaten the function of the brain since it prevents proper air intake into your lungs.* You know the burden of carrying extra weight in your daily activities, from stair-climbing and getting into and out of a car. When you slim down on the ECD Program, you protect your glands, are able to breathe better, can think better, get around more youthfully. Your entire personality is rewarded with the same vitality as your body, when it is slim-trim under the ECD Program.

HOW ENZYME DESSERTS CAN HELP SLIM
AN EXPANDING WAISTLINE

Caroline B. was troubled with her "middle-aged spread." She was embarrassed because she was not even middle-aged! In a desperate effort, she gave up her desserts. But she still kept gaining. An understanding neighbor told her that desserts—using everyday fruits—could actually help whittle away her waistline. She could enjoy her desserts while slimming down.

Simple Enzyme Program: With each of her meals, Caroline B. had a raw fruit salad. The enzymes in the raw fruit worked to metabolize accumulated fats, carbohydrates and calories. The weight loss was created by the way her body made use of the fruit enzymes.

Fruit enzymes work speedily to improve better metabolism and assimilation and excretion. Fruit enzymes helped burn the calories rather than letting them become stored. Without enzymes, fats and carbohydrates store calories without burning a substantial amount to create energy. By eating a fruit dessert, the enzymes create the needed combustion so that the cells can be "washed" through assimilation.

Raw Fruits Are Key To Waistline Whittling. Enzymes live in *raw* fruits. To promote the needed metabolism, you need to eat *raw* fruits. Processed foods are usually devoid of enzymes since excessive heat and cold destroy them, as does storage. The key to enzyme catalyst action is to eat *raw* fruits.

Become Slim, Trim, Healthful. Within one week, Caroline B. had whittled some two inches from her waistline, thanks to the use of fresh, raw fruit as a dessert with each of her meals. Now she no longer looked or felt middle-aged. She was slim, trim and healthful. She radiated the same youthful vitality in her attitude and appearance.

THE BASIC 13-STEP ENZYME CATALYST DIET PROGRAM FOR SLIMMING DOWN

To alert your weight-melting metabolism, enzymes need a proper environment. When enzymes promote better internal combustion, your billions of body cells are constantly being washed, so that excess fat cannot be accumulated. The following basic 13-step program builds up enzyme content and activity in your body. This program helps promote better metabolism to "burn off" unwanted pounds. It can be followed easily right in your own home.

1. Avoid late, heavy dinners. Whenever possible, eat your major meal at midday; if this is impossible, have dinner as early in the day as you are able to manage. **ECD Benefit**: You avoid overloading your enzymes with a heavy night meal, immediately followed by sleep. Food eaten earlier in the day, when you are more physically active, can be better digested by your enzymes. Metabolism is more vigorous during the early part of your day when your muscles are more active in bodily motions.

2. Eat foods slowly. Always chew food thoroughly; do not bolt it down. **ECD Benefit**: Slow chewing brings forth a greater supply of mouth and digestive enzymes. These catalysts can melt accumulated fat from your cells much more efficiently if the foods have been carefully, slowly, and thoroughly chewed. Finely pulverized foods also protect against gassy indigestion caused by improper eating or swallowing.

3. Drink natural beverages. Avoid coffee, tea, cola drinks, bubbly drinks. **ECD Benefit**: Your body will have full enzyme strength to melt down fat. Beverages containing caffeine are destructive to enzymes. Constant intake of caffeine will weaken and destroy enzyme power, so weight builds up. Furthermore, caffeine in these beverages tends to redden the gastric mucosa and reduce the fat-melting strength of these catalysts. Instead, switch to healthful fruit and vegetable juices, which are prime sources of enzymes. Try

coffee substitutes such as Postum. Drink herbal teas. Sweeten with a bit of honey for flavor.

4. Avoid harsh spices and condiments such as pepper, mustard, vinegar, catsup, pickles, salt. **ECD Benefit:** The enzymes found in your alimentary tract as well as in your liver and kidneys will remain at full catalytic strength. Volatile substances like harsh spices inhibit enzyme catalyst action, thereby allowing more weight buildup. Harsh condiments also destroy enzymes themselves throughout the digestive system. Rather than harsh spices, try flavorful herbs for seasonings to soothe your enzymes.

5. For a salad dressing, use high-enzyme raw fruit juices instead of harsh vinegar. **ECD Benefit:** Fruit juices are prime sources of fat-dissolving enzymes. They join with the enzymes of a raw vegetable salad and give a catalyst action that will help slim down your fat cells. Furthermore, harsh vinegar contains acetic acid. Concentrated, it can remove warts and moles. But when consumed, it destroys liver enzymes. It weakens your general metabolism so that the catalyst action of enzymes is reduced and even destroyed. Try any raw fruit juice with a bit of polyunsaturated oil for a delicious enzyme-soothing dressing.

6. If vegetables must be cooked, do so until barely done. Eat raw vegetables as often as possible. **ECD Benefit:** Raw vegetables are prime sources of fat-melting catalysts. Cooking can weaken these enzymes or destroy them at high heat. A rule of thumb is to cook vegetables in as little water as possible for under ten minutes, unless more time is absolutely required. This preserves the needed enzymes that will help melt the fat from the body cells.

7. For unique enzyme power, eat highly acid fruits at separate intervals from coarse, fibrous vegetables. **ECD Benefit:** The metabolism recharging action of enzymes remains at full power. This recharging action may become sluggish if *both* fruits and vegetables are consumed at the same time. Fruits require almost no digestion. If they must be held in your digestive system until the vegetables are ready to pass through, there may be fermentation. This weakens and slows the catalyst vigor of enzymes. They cannot perform their cell-slimming action. Therefore, the overweight person would do wise to eat fruits and vegetables at separate meals when beginning the ECD Program. Gradually, as weight is lost, they can be eaten in combination, on occasion.

8. Avoid custards or rich puddings containing milk, eggs, and cream with sugar combined. This rich mixture, used in large amounts

with eggs, tends to coat the digestive tract, often "sealing" the cells so that enzymes cannot penetrate to the nucleus to create the fat-melting catalyst action. Drink milk alone. Give up any form of refined sugar. Eat a maximum of three eggs weekly. When you finish either a milk or an egg food, top it off with any raw fruit. **ECD Benefit**: Enzymes will then be able to promote better digestion and be able to penetrate the cell for better reducing.

9. Foods should be baked, broiled or boiled. Avoid fried foods in any form. Fried foods are coated with grease which "locks out" enzymes. This creates a sluggish digestion. Cells become "engorged" and fat. Enzymes cannot reach the cells because of the thick fried fat coating and weight can accumulate. **ECD Benefit**: Foods that are baked, broiled or boiled have no such "hard core seal" and enzymes can penetrate them and perform the cell-slimming catalyst action.

10. Avoid heavy (or any) liquid intake with your meals. Large amounts of water dilute your digestive enzymes and inhibit their catalyst actions. Drink freely *between* meals. Drink lots of raw fruit and vegetable juices *between* meals for enzyme vigor. Too much liquid consumed *with* meals can "drown" enzymes. Overly-hydrated cells tend to absorb calories and carbohydrates and fats. The catalyst action of enzymes is diluted and watered down when too many liquids are consumed with meals and the cells accumulate weight. If you must drink, then sip a fruit or vegetable juice. Otherwise, wait at least 60 minutes after a meal for your beverage. **ECD Benefit**: Your enzymes will remain at undiluted strength.

11. Avoid processed or synthetic "cheese foods". These cheese foods contain enzyme-destroying chemicals. In the ripening of processed cheese, certain putrefactive bacteria develop which destroy enzymes. You may, however, use cottage cheese freely. Cream cheese is also heal.thful. **ECD Benefit**: Both these forms of cheese contain lactic-acid bacteria, which protect against putrefaction and appear to soothe the enzymatic system. These cheeses are healthful for the enzymes and prompt them to issue forth a strong catalyst action, helping to slough away fat deposits from the cells and promote slimming down.

12. Select whole grain and natural flour products, rather than refined or processed foods like white bread, sugar, polished rice and all bleached white-flour products such as spaghetti and macaroni, etc. **ECD Benefit**: Enzymes do not have to deal with bleaching chemical additives, which can destroy them. Many overweight people are huge devourers of refined and bleached foods, which are both enzyme-

dead and deadly to enzymes . . . hence the vicious cycle of over-weight! Instead, select whole grain and natural flour products, prime sources of B-complex vitamins used by enzymes to build up better catalyst-metabolic action and cell-slimming processes.

13. Keep meals simple, but attractive. Try to minimize complicated mixtures and reduce too many varieties at the same time. **ECD Benefit:** Enzymes can promote a better catalyst action if not diverted by having to work with so many varieties of foods. As your weight goes off, you can gradually add more varieties, if you must have them. But at the start of your ECD Program, keep varieties down to a tasty simplicity.

More Tips: Try not to eat between meals since this overburdens your enzymes and weakens their weight-melting powers. Your digestive system should empty one meal, have a rest period before taking on another batch of food. Enzymes need "rest" as well as your other body organs and systems. Unless your digestive system is given this rest, it will become like a milk crock that is not cleaned between milkings. The enzyme system goes sour. Weight builds up. Instead, let your digestive system rest for better catalyst action. Drink more fruit and vegetable juices to further boost their catalyst vigor.

The Enzyme-Catalyst Diet gives your body the raw working materials with which fat cells can be reduced and weight can then be lost. It is the natural, effective and safe way to slim down . . . once and for all!

EIGHT ECD REWARDS THAT HELP PROMOTE HEALTH WHILE SLIMMING YOU DOWN

What makes the Enzyme-Catalyst Diet Program so unique? Unlike other diet programs that call for restrictions or denials or risky drugs or medications, it helps promote your health while slimming you down. Here are eight ECD rewards that you enjoy on this slimming program:

1. *Feel Energetic While Slimming.* The ECD Program supplies you lots of energy. It is rich in amino acids, polyunsaturated oils, vitamins, yet it is low in hard fats that are the cause of overweight. The ECD Program gives you sustained energy with minimal fluctuation.

2. *Be Nourished With Minerals.* The ECD Program gives you a balanced calcium-phosphorus and magnesium ratio. It is a prime source of minerals, needed by your body when you slim down.

Enzymes use these minerals to protect you against infection and enrich your bloodstream as pounds are melted away.

3. *All-Natural, Non-Toxic.* The ECD Program has no harmful ingredients, no preservatives, no synthetics, no spray residues, no refined sugar, no hard fats, no bleached flours. It is non-toxic. It is all-natural!

4. *Melts Fat The Natural Way.* The ECD Program calls for using enzymes found in raw foods to digest, metabolize and then wash away fat from the billions of body cells. This is the natural way to melt away fat and create youthful slimming.

5. *Non-Habit Forming.* The ECD Program uses no drugs, no medications, no medicines. It is non-habit forming. It uses *only* healthful foods you can find in your corner market, in a proper combination.

6. *Non-Constipating.* Many diet programs create constipation because of severe restrictions. The ECD Program calls for the use of enzyme foods which give you a soft, natural stool. There is no need for chemical stool softeners.

7. *Slims Waistline.* Much weight is found along the waistline, hips and thighs. This area, a "fat cell zone" houses your digestive and intestinal tract. Here, fat is "born." The ECD Program uses natural food enzymes to promote a natural shrinkage of your intestinal tract, so that your waistline can then slim down.

8. *Improves Digestion.* The very foundation of the ECD Program is improved digestion so that enzymes can "attack" accumulated fats, carbohydrates and calories and metabolize them, dispose of them to help maintain a healthful weight.

The Enzyme-Catalyst Diet Program lets you enjoy good, healthful foods. No need to starve yourself. No need to fear drugs which have reactions and side effects. The ECD Program is not a torture regimen. It does not make you feel weak. It can help you lose much more than you ever thought you could. It works on the principle of controlling your body's "furnace" or metabolism so that with the alerting of "body combustion," your accumulated fat in your cells are enzyme-digested and disposed off. You are rewarded with a new slim-trim *You* that everyone will admire and love . . . including yourself!

Highlights:

1. Feast your way to slimming with the Enzyme-Catalyst Diet

Program. No need to deprive yourself of most foods. Let enzymes do your reducing!

2. Set a desired and healthful weight goal and then weigh yourself regularly according to the charts.

3. The ECD Program reduces your "fat" cells, thereby promoting slimming from its very source ... namely, your overweight cells.

4. Enjoy sex life and health building benefits, in addition to weight loss, with the ECD Program.

5. Caroline B. melted down her "middle-age spread" by eating desserts!

6. Let enzymes keep you slim, by using the 13-step ECD Program in daily living. Easy to follow. Tasty, too.

7. The ECD Program has eight unique rewards for the happy dieter.

2

How the Enzyme-Catalyst Diet (ECD) Melts Overweight Pounds

A delicate life-like substance created by Nature and found almost exclusively in *raw foods* holds the key to melting away overweight and unwanted pounds. The name of the substance, *enzyme,* is taken from the Greek word *enzymos,* which means "fermented" or "leavened"–to create a catalyst or change. Raw food enzymes are catalysts because of their ability to stimulate an internal reaction without themselves being transformed and destroyed in the process. Simply speaking, and enzyme is an internal fluid or sap created by the living tissues and cells of the digestive tract. They are secreted by the digestive tract and then set to digest or metabolize eaten foods, to alert metabolism and to set off the act of catabolism wherein foods are broken down, assimilated and the excess cast out of the body.

Enzymes Are Spark Of Life. Enzymes digest all of the food to make it small enough to pass through the tiny pores of the intestines, into the bloodstream. Life could not exist without enzymes. They help rebuild the prepared food into muscle, nerve, bone, gland. Enzymes help store excess food in the liver or muscles for future use. Enzymes participate in creating the formation of urea to be eliminated in the wastes, and also in the elimination of carbon dioxide from the lungs.

Nourish Bloodstream, Promotes Oxidation, Melts Pounds. Enzymes help build phosphorous into bone, blood and nerve. Enzymes help fix iron in the red blood cells. Enzymes are used to metabolize substances in food to build this iron structure in the bloodstream. Enzymes cause coagulation of the blood and stop bleeding. Enzymes decompose toxic hydrogen peroxide and liberate healthful oxygen from it. Enzymes promote oxidation, the union of oxygen with other substances. Enzymes attack waste materials in the blood and tissues and transform them into substances that can be eliminated through the waste channels of the body. Enzymes change protein into fat or sugar. Enzymes change carbohydrates into fat. Enzymes help burn up calories to produce energy. Enzymes hold the key to the melting of overweight pounds.

THE REAL REASON WHY YOU ARE FAT . . . AND HOW THE ECD PROGRAM CAN MAKE YOU SLIM FOREVER

The Fat Is In Your Cells. Microscopically, your billions of cells resemble little gray globs enclosed in thin sheaths. The gray contains tiny specks; the heart of the cell is a darker gray glob (the nucleus), also sealed within a thin sheath. Ordinarily, a thousand such cells, lined up, would cross the head of a pin. But millions of them can add up to a lot of weight. Fat cells (called *adipocytes*) are alive. They participate in most body processes, requiring oxygen, blood circulation, fuel and protein building blocks.

But these living fat cells, or *adipocytes,* are also *storage tanks.* The fat that is to be stored is transported to the middle of the cell. The gray cytoplasm and all the cell's working components, are pushed out to the border in what is known as a "signet-ring" shape. The nucleus is in the signet part. The yellow fat now becomes the core of the cell. When the body has a deficiency of raw food enzymes, these fat cells become engorged. *Adipocytes* become enlarged, overpacked. Over a period of time, these fat cells become so heavy, the problem of overweight is well underway.

Size And Not Amount Of Fat Cells Cause Overweight. The *size* of your *adipocytes* or fat cells cause overweight and *not* the amount of fat cells that you have since puberty. Even if you lose weight on a reduced food intake, you still have the same amount of fat cells. When you start eating heavily, the fat cells start getting heavier and weight increases. The little-known secret of fast, permanent weight loss is to *keep your fat cells from increasing in size.* The use of raw food enzymes is the most effective way of keeping your fat cells slim . . . and your *body* is then kept naturally slim. In brief, *you are as slim as your cells!*

HOW THE ECD PROGRAM MELTED 30 POUNDS IN 30 DAYS

Marge K. was always tasting food that she prepared for her family. This tempting habit put more than 30 pounds on her hips, abdomen, thighs, even her legs. Marge K. had tried diet pills but these caused her to break out in a skin rash; she had fits of dizziness and emotional upheaval. She had to give them up. She needed a natural way of helping to melt away the increasing pounds that kept building up in a frightening way. Weighing herself daily became a shameful experience. She kept gaining and gaining. She looked it, too, with bulges in her stomach and spreading hips. She even developed an unsightly double chin. A "suddenly slim" relative came for a visit and told how she had been to a "health farm" where raw foods were served, almost exclusively. She had lost some 20 pounds within two weeks. She told Marge K., the basics of this *Enzyme-Catalyst Diet Program:*

1. All fruits should be eaten raw. While fresh fruits are prime sources of enzymes, frozen may be used if fresh is unavailable. (Try to avoid canned fruits since these have been heated and enzymes either depleted or destroyed.) Eat an assortment of raw fruits *before* any meal. Eat fresh raw fruit throughout the day whenever you want to have a snack.

2. All vegetables should be eaten raw. Vary your salads, using different vegetables with your meals. Try a green salad with sliced tomatoes, carrot sticks, celery. For dressing, use a little apple cider vinegar and oil with a sprinkle of lemon juice and honey.

3. Begin each and every meal with a raw food. *This is important!* The raw food enzymes thus prepare the body for the food that will soon enter and lie in waiting to metabolize it to help prevent against fat buildup.

Results: Marge K. followed this plan for 30 days. She weighed herself daily and cheerfully noted that the scale showed she was losing overweight pounds. Before the 30 days were over, she had shed the unsightly excess weight. Now she follows through by *keeping off the weight* by using the 3-step ECD Program as outlined above. She is "suddenly slim" and "suddenly youthful" thanks to enzymes.

Secret Of Fat-Melting Power of ECD: Raw food enzymes gather together to wait for food that is to be ingested. They participate in alerting metabolism so that when fats, carbohydrates and calories enter the digestive system, there is a boost in activity. The enzymes enter within the nucleus of the fat cells or *adipocytes.* They actually penetrate the gray cytoplasm, melt down the accumulated fats, carbohydrates and calories and, literally, burn them right out of your body. This is known as metabolism and assimilation. *The raw food enzymes perform this pound melting action.* Without raw food enzymes, or if there is a deficiency, the *adipocytes* become enlarged, engorged and excessively fat. This causes overweight. Protection against fat cell enlargement and overweight is possible with raw food enzymes!

HOW A FOOD LOVER COULD FILL UP . . .
BUT NOT FATTEN UP . . . WITH ENZYMES

Daniel R. loved good food. Raised on a farm where the groaning table was a tradition, Daniel R. could not easily give up his childhood passion for good food. There is nothing wrong with good food . . . except when you eat too much of it! So it was with Daniel R. He scoffed at being called "fat" but agreed to try not one but three different "at home" tests to be convinced that he was overweight:

1. *Pinch Test.* Pinch the back of your upper arm between your tumb and first finger. If this fold of skin and fat is much greater than an inch, you are excessively fat.

2. *Ruler Test.* Lie down on your back. If a ruler resting lengthwise along the middle of your abdomen will touch both your ribs and pelvic area, you are not too fat. If it cannot touch, then you are too fat.

3. *Beltline Test.* If your beltline is longer than the circumference of your chest at the nipples, you have too much abdominal fat.

Daniel R. took all three tests. He failed on all of them. Convinced he was too fat, he agreed to follow the ECD Program. He had tried other reducing programs but they denied him much delicious food. He wanted to eat and lose. With the ECD Program, this was possible. Here is the basic 3-step program followed by food-loving Daniel R.

1. Super-Active Enzymes For High Fat Foods

Before eating any high fat food such as meat, poultry, fried foods, gravy saturated foods, eat a bowl or a plate-full of these super-active raw enzyme foods: cranberries, gooseberries, grapefruits, currants, kumquats, lemon wedges, lime wedges, loganberries, oranges, pineapple slices, pomegranates, sour grapes, strawberries, tangerines, tomatoes, radishes, sauerkraut, watercress.

2. Active Enzymes For High Carbohydrate Foods

Before eating any high carbohydrate food such as spaghetti, macaroni, baked grains, puddings, beans, noodles, oatmeal, pancakes, salad dressings, eat a bowl or a plate-full of these active raw enzyme foods: apple, apricot, blackberry, cherries, elderberries, grapes, guava, huckleberries, mango, nectarine, papaya, peaches, pears, persimmons, plums, raspberries, lettuce, seasonal melons, cantaloupe, onion rings in a salad.

3. Gentle Enzymes For High Calorie Foods

Before eating any high calorie food such as desserts, breads and foods made with bread, gravies and drippings, breakfast cereals, confections, any food to which sugar or a sweetening has been added, eat a bowl or a plate-full of these gentle raw enzyme foods: bananas, dates, figs, Thompson and Muscat grapes, prunes, raisins, raw seeds and raw nuts, almonds, celery, carrots, cucumbers, cabbage, raisins, watermelon, raw wheat germ, sesame seeds.

How This 3-Step ECD Program Helped Daniel R. Fill Up ... But Not Fatten Out. Close to 1000 enzymes are known to exist. There are strong enzymes that act upon strong foods. There are mild enzymes that act upon mild foods. When Daniel R. ate the listed raw food enzymes *before* partaking of either the fat, carbohydrate or calorie food, his digestive system used these enzymes to metabolize the specific type of food being metabolized. The "high fat enzymes" were especially strong in penetrating the thick fat within the cell nucleus and metabolize it to guard against buildup. The "high carbohydrate enzymes" were not as powerful as the fat enzymes, and were used to metabolize less-powerful carbohydrate foods. Finally, the gentle enzymes were used for "high calorie foods" which require less vigorous power for metabolism and assimilation.

Created "Sponge-Like" Slimming Action. The reason for overweight lies in your body's fat cells. Just as a sponge with many cells will soak up more water and hence become heavier, so will your billions of fat cells become engorged and heavy if pound-causing fats, carbohydrates and calories are allowed to accumulate. Just as you can squeeze water out of a sponge and "reduce" it in weight, so do enzymes actually squeeze or "melt" down excess fats, carbohydrates and calories within the nucleus of your fat cells or *adipocytes,* and thereby "reduce" weight.

When Daniel R. *began* with a raw enzyme food, he sent these metabolic miracle workers shooting throughout his digestive system where they helped metabolize and "burn up" pound-causing substances. It was the natural way for him to enjoy good food and keep slim. He soon lost close to 47 pounds. Now, his skin folds were slim, his waist was lean. He looked and felt good, thanks to the ECD Program.

ECD Program Reduces Body By Reducing Fat Cells. Raw food enzymes control fat buildup in the fat cells which appear to lie in wait, ready to become engorged with pound-building fats. Raw food enzymes transform fats, carbohydrates, calories and other weight gainers into substances that can be, literally, washed out of the body. The ECD Program improves the health of the kidneys to help them dispose of more fluids, thereby guarding against edema (water retention) of the cells and excess weight. The ECD Program improves the cellular ability to burn starches and sugars and to control how much will be stored as fat. The ECD Program keeps the body slim and trim . . . while you enjoy most of your favorite foods.

HOW TO ALERT-ACTIVATE-AMPLIFY
YOUR OWN ENZYME NETWORK

Create a healthy *enzyme environment* wherein your digestive processes will work in smooth harmony. In so doing, you will alert your sluggish enzymes, activate your natural enzymes, amplify your enzyme fat-melting activities. Here is a step-by-step program that improves the function of enzymes:

1. Do not eat standing up. This causes a constriction of digestive muscles and inhibits the effectiveness of enzymes to attack excessive fat in your *adipocytes* or fat cells. Eat only when sitting down, and completely comfortable in body and mind.

2. Eat in the same place regularly. This prepares your digestive system. Your enzymatic network feels "secure" and "familiar" and this helps amplify their fat-melting methods. It also

helps cut down on excessive eating. Train yourself to eat only in your kitchen or dining room. This cuts out eating before the TV set or sipping a beer in your back yard.

3. When eating, devote full concentration to your food. Do not read or watch television while eating. Emotionally, focus your attentions on your food. This helps create a "happy enzyme" environment and digestive processes are improved.

4. Use a smaller plate. A smaller portion will look larger if put on a smaller plate. This also eases your urge for excessive eating. Your enzymes will be content with a smaller portion if, visually, you are satisfied.

5. Chew your food carefully. Your salivary glands contain enzymes which begin digesting your food and changing them into a form that can be accommodated by your body cells. Chewing activates mouth enzymes, alerts digestive enzymes to the food that will soon be received and helps amplify their function. Proper chewing releases enzymes that help pre-digest food and control fat buildup in your *adipocytes.*

6. Take smaller bites. Put your fork down between each bite. This slows your eating pace and gives your enzyme system an awareness of being nourished. It creates better metabolism and fat-burning results.

7. Eliminate unhealthful beverages. These include coffee and tea, which contain caffeine. Caffeine lowers blood sugar, creates hunger, and also interferes with the natural digestive power of your enzymes. Avoid so-called "diet cola" beverages. Not only do they contain artificial sweeteners which are chemical antagonists and destructive to enzymes, but they also pack plenty of caffeine. Switch to enzyme-rich fruit and vegetable juices.

8. Ask others to help you. When you are motivated, your emotions alert your enzymes and they can perform more efficiently. Seek their encouragement and help. It puts you in a good emotional state of mind and this creates better enzymatic activity.

9. Boost protein for appetite satisfaction. Your enzymatic system can take care of protein easier. Carbohydrates and fats are absorbed more speedily and solidly by the nucleus

of your *adipocytes* or fat cells. Enzymes will require more effort to wash them out of your cells. Protein intake should be higher than carbohydrates or fats.

10. When eating, your clothes should be comfortable but *not* restricting. A tight belt or corset can "choke" your enzymes so digestive powers are weakened and fat cells become swelled up.

HOW THE ECD PROGRAM
OFFERS PERMANENT WEIGHT LOSS

The Enzyme-Catalyst Diet Program succeeds where other conventional reducing diets often fail. The reason here is that most diets call for restricted carbohydrates or calories or fats, or various combinations. But the *fat cells* are still allowed to become engorged with whatever heavy ingredients can enter the nucleus and remain. Since your billions of fat cells are comparable to a sponge, they can absorb quite a bit. Something has to "squeeze" out the weight from this sponge. Enzymes perform this task by getting to the *cause* of the overweight . . . in the fat cells. Other diets may cause weight loss, but also create what is known as a "Yo-Yo Syndrome" or a "See-Saw" Weight. First it's up, then it's down, then it's up again!

How The ECD Diet Program Avoids "See-Saw" Weight. This so-called weight loss is cyclical. You lose weight, but then regain it . . . and then some. Your weight goes up and down repeatedly, like a see-saw. But the ECD Diet Program does more than help you shed fats, carbohydrates and calories. It goes to your *adipocytes* and actually "burns up" the weight. This makes you as permanently slim as your cells! Conventional diets do not always slim down the cells, hence the "See-Saw" weight fluctuations.

HOW JENNY S. USED THE ECD PROGRAM
FOR FAST "FOREVER-SLIM" RESULTS

Jenny S. used to say that whatever she ate turned to fat. She was, indeed, corpulent. She had thick arms, a heavy, sagging chin, dropping breasts, a thick "spare tire" around her middle, and unsightly clumps of thighs. Even her calves were heavy. She looked much older (and felt it, too) than her middle 40's. But Jenny S. said she was not to blame. She tried one diet after another. She did lose weight. But the moment she stopped the diet, she gained it back again . . . and much more. She was a victim of "See-Saw" weight.

How ECD Program Corrects "See-Saw" Weight Problem: When Jenny S. would overeat, she would weaken or tire her metabolic capacity; her enzymes could not manage this huge intake of food. When she reduced food intake, she ate so sparingly, she was deficient in enzymes and felt sick and weak. This zig-zag practice caused her weight to go up, then down, then up again.

Enzymes in raw foods were needed, together with a simple diet program. Here is what Jenny S. was told to do:

Raw Fruits Daily. Three times daily, eat an assortment of seasonal raw fruits. Citrus fruits (oranges, grapefruits, lemon and lime wedges, nectarines) are a *must* for a strong enzyme content.

Raw Salads Daily. Eat as many as you desire. Slice raw celery, radishes, cucumbers, carrots, chicory, escarole, mushrooms, turnips. A raw salad is a powerhouse of enzymes. Eat several, daily.

Green Leafy Vegetables. Also include yellow vegetables such as squash, pumpkin, sweet potatoes. Eat cabbage, string beans, asparagus regularly. Steam in a little boiling water, until just tender, for good enzyme power.

Protein Foods. Lean meats, poultry, fish, cheese and eggs. Always trim fat from all meats. Avoid fatty meats such as pork, duck, sweetbreads, ham. These require very tough enzymatic fat-melting and divert enzymes from other foods so you still gain weight. (Meat and fish should be broiled or baked for better digestive action.)

Mineral Foods. These include green and yellow vegetables, as well as cottage cheese, skimmed buttermilk, fat-free skimmed milk, skimmed milk, yogurt. All are prime sources of minerals and enzymes and help improve their fat-melting powers.

Jenny S. Loses 58 Unsightly, Unhealthy Pounds. This 5-step daily eating ECD Program boosted Jenny's enzyme power. Now, the enzymes could attack the fat in her *adipocytes* and get to the *cause* of her overweight. Now she could be more than just slim. She could be *permanently slim.* The enzymes squeezed, so to speak, the fat out of her cells and kept her slim. She just had to eat delicious raw fruits and vegetables daily. Soon, she lost some 58 pounds. No longer was she the victim of on-again, off-again weight, or the "See-Saw" weight problem. The weight was gone . . . and stayed gone!

YOUR DAILY ENZYME-CATALYST DIET EATING PROGRAM

To give your body cell-washing enzymes, here is a basic program. Note that it lets you eat a variety of your favorite foods. But you

MUST eat the raw foods daily, so that you can take off weight and keep it off.

EAT THIS WAY EVERY DAY

1. Dark Green Leafy and Deep Yellow Vegetables

Choose: Broccoli, carrots, chicory, escarole, greens (beet, collard, dandelion, mustard and turnip), kale, pumpkin, sweet potatoes and yams, Swiss chard, watercress, winter squash, apricots, cantaloupe, mango, papaya.

How Much: One or more servings every day. Select fresh or frozen. Good source of Vitamin A, iron and Vitamin C along with powerful enzymes. Other good sources include asparagus, snap beans, Brussels sprouts, green peas, peppers, peaches.

2. Whole Grain Bread and Cereals

Choose: Whole grain bread, rolls, cereals, flour, cornmeal, brown rice, grits.

How Much: One or two servings per day. (One half cup cereal equals one slice bread.) Limit or eliminate cakes, cookies and foods made from commercial mixes. They antagonize and chemically destroy enzymes.

3. Citrus Fruits and Other Plant Foods

Choose: Grapefruit, oranges, tangerines, raw cabbage, cantaloupe, strawberries, guava, mango, papaya, tomatoes. Also fresh juices made from these plant foods.

How Much: Three or more servings per day. Select fresh or frozen. Good source of Vitamin C as well as strong enzymes.

4. Dairy Products

Choose: Skim milk, buttermilk, evaporated skimmed milk, plain yogurt, non-fat dry milk for drinking, cooking, baking, cottage and pot cheeses.

How Much: 2 cups for most adults. A prime source of calcium, riboflavin and protein. These nutrients become part of your fat cells, and help shield against excessive fat buildup. They are also used by enzymes for strength and vigor in fighting fat.

5. Meat, Poultry, Fish, Eggs, Dried Beans and Peas, Nuts

Choose: Poultry and lean meat. Fat or lean fish. 2 to 3 eggs per week. Dried beans, peas, nuts.

How Much: 1 serving per day. Emphasize fish which offers a rich source of polyunsaturated fatty acids which are taken up by enzymes to nourish the *adipocytes*. Enzymes will also use this protein as a protective element against hard fat invasion of the cells.

6. Fats

Choose: Corn, soybean or safflower oils for cooking, baking and salads. Margarines containing significant amounts of one or more of these three oils in *liquid* form.

How Much: One ounce (two tablespoons) of oil every day. Enzymes use these polyunsaturated fats to offer you satisfaction, to moisten and lubricate your *adipocytes* and help "flush out" the hard fats and help keep your cells (and you, too) permanently slim.

Sample Enzyme Catalyst Diet Eating Plan	
BREAKFAST	LUNCH

Raw fruit salad and/or juice
Protein food (milk, or egg, or
    cottage cheese or lean
    fish.)
 	
Whole grain bread or cereal
Hot beverage
	Raw vegetable salad
	Hot dish or sandwich
	    of cheese, raw
	    vegetables, lean
	    meat, poultry
	Raw vegetable juice
	Season fruit salad

DINNER	SNACKS
Cooked beans or protein food Baked potato Dark green leafy and deep yellow vegetable salad Raw fruit salad Beverage	Skim milk Fruit or vegetable juices Assorted seeds, nuts Crisp vegetable chunks Assorted fruits

EAT "LIVING" FOODS FOR "LIVING" ENZYMES

Most foods should be eaten fresh and uncooked. Note that heat, such as cooking, can destroy food enzymes. Anything higher than 120°F. will destroy the enzymes in the food. Your fat cells are thus deprived of their powers in metabolizing fats, carbohydrates and calories. Now, quite realistically, it is unsavory and unhealthy to eat raw or uncooked meats (they contain bacterial parasites) or raw eggs, fish, poultry, beans, etc.

But you can select foods to be eaten with cooked meats or other proteins, foods which can be eaten raw and which are "alive" with "alive" enzymes.

Daily, include unprocessed or raw wheat kernels (available at special diet shops and health food stores), certified raw cow's milk wherever available, fresh raw fruits and vegetables, natural raw fruit and vegetable juices. Build these enzyme foods into your daily eating programs. They help give you good nutrition, satisfy your tastes, protect you from between-meal hunger, give you a sense of well-being, do not make you feel tired.

Most important, the Enzyme-Catalyst Diet will also get to the root-cause of your overweight. Namely, the billions of fat cells throughout your body, lying in wait to absorb materials that swell them up like a sponge and fill you out so that you become fat! With enzymes, you offer your billions of cells the metabolic power to dissolve excess fat that will be disposed of . . . and keep you permanently slim. Enzymes, therefore, correct the *cause* of over-weight, and this is the secret of the success of the ECD Program.

In Review:

1. The fat is in your *adipocytes* or body cells. Use enzymes to dissolve excess cellular fat and you can become slim forever.

2. A simple 3-step basic ECD Program helps slim down your cells.

3. Marge K. took weight off her body (30 pounds in 30 days) under the tasty ECD Program.

4. Daniel R., a food lover, could fill up, but not fatten up, with enzymes. He began each meal with various enzyme foods, found everywhere, and created quick slimming action. He lost 47 pounds in no time at all.

5. You can alert-activate-amplify your own enzyme network in 10 easy steps. Takes minutes. Works wonders for a lifetime of slimness.

6. The ECD Diet Program guards against "See-Saw" Weight by taking it off permanently . . . while you enjoy favorite foods.

7. Jenny S. disposed of her thick arms, sagging skin, "spare tire" under the tasty ECD Program. More important, it was permanent weight loss.

8. Eat wholesome, tasty foods, every day, for fast and "forever slim" benefits.

3

How to Use the ECD Program to Wash Out 10-20-40 Pounds of Unwanted Weight

Within your body are "weight-melting" enzymes that can easily be activated to wash out 10-20-40 or more pounds of unwanted weight. *Enzymes,* do more than help you lose that excess fat. These all-natural pound-dissolvers *keep off that excess weight!* Once these catalysts are alerted, they become "built-in" body weight controls. They let you eat sufficiently, enjoy most of your favorite foods, but help keep your weight down to a healthful level.

"Weight-Melting" Power Of ECD. The enzyme-catalyst diet breaks down food that you eat, penetrates the fat cells where fats, carbohydrates and calories are stored, and promotes internal metabolism. This action metabolizes accumulated "weight" and helps you slim down almost overnight. To alert these enzymes, you need to make some adjustments in *the way you eat,* as well as what you eat. Once you have alerted these enzymes, they immediately work to catalyze or dissolve weight-causing substances.

HOW TO CHEW YOUR WAY TO WEIGHT LOSS

Before you take a mouthful of food, take a moment to *think* about the food. Take another moment to *study* the food. Then, slowly eat the food. Once you have taken the food into your mouth, *chew it thoroughly.*

ECD Weight-Melting Action: The process of metabolism is alerted as soon as you decide to eat a certain food. Your mental attitude activates weight-melting enzymes. Once you start to chew, *salivary enzymes* begin to metabolize the food. An important weight-melting enzyme in your mouth, *ptyalin,* begins by quickly attacking the carbohydrates of bread and other starchy foods, breaking them down into *maltose,* a predigested form of starch, or energy-producing sugar. This action, begun by your mouth enzymes, break down carbohydrates into a form that is more easily metabolized. *Less carbohydrates build up as fat, if you take the time to chew your food very thoroughly!* Just two or three minutes of chewing can help you lose up to 40 or even 50 pounds of unwanted weight before the month is over. You can therefore enjoy carbohydrate foods without gaining unnecessary weight, with thorough chewing.

HOW TO DIGEST YOUR FOOD FOR LESS WEIGHT BUILD UP

After you swallow your food, new enzymes poured forth by the digestive glands are made ready to greet the arrival. NOTE: As you chew, important weight-melting digestive enzymes are poured forth in readiness. If you bolt down your food, it is improperly broken down by mouth enzymes; in your stomach, weak or insufficient enzymes cannot properly metabolize them and they build up in your fat cells. Your body has ready-to-work enzymes if you alert them properly. When you swallow your food, both during the meal and even afterwards, give yourself a 30 minute rest period for better enzyme activity.

ECD Weight-Melting Action: If you enjoy milk foods, you can indulge but you need to give your digestive enzymes time to metabolize them to protect against weight build-up. The *rennin* enzymes create a catalyst action to coagulate the milk protein and change it into *casein,* in a form that can be better metabolized by your body. The *rennin* enzyme also metabolizes ingredients in cheeses so they can be used to build body health and not stuff the fat cells. Two other enzymes, *pepsin* and *lipase,* work to digest protein and fats, splitting them into forms that can be assimilated by your body and carried through your bloodstream to strengthen your bones, nerves, teeth and other parts. These enzymes act as a catalyst so that protein, fats, carbohydrates, calories are *used* for body health, rather than excessively *stored* as body fat!

You need to give your digestive enzymes more time to create this action. *A basic rule of thumb is to chew thoroughly, eat slowly, and then rest after your meal.* Enzymes perform a catalyst weight-melting action when your digestive system is relaxed and not subjected to constricting influences.

HOW TO USE A POWERFUL BODY ENZYME
FOR QUICK WEIGHT MELTING

A most vital digestive enzyme is *hydrochloric acid.* It is so powerful that if the stomach were without a protective coating, it could burn right through the lining. Nature provided you with this enzyme to attack *hard fats* and *strong carbohydrate foods,* to break through these foods and metabolize them so they can be used, and not stored, by your body. To alert a sluggish hydrochloric acid flow, eat one or two freshly washed raw apples *before* eating a heavier meal.

ECD Weight-Melting Action: Once you have alerted your digestive glands, through raw apples (good source of fruit enzymes), they begin to secrete hydrochloric acid. This enzyme acts on strong protein foods such as meats, fish, eggs, dairy foods. This enzyme then acts upon tough fibrous vegetable cells (it works on the bran portion of wheat to take out needed grain vitamins, minerals, proteins, carbohydrates and protect against weight build-up through metabolism) and also regulates the delicate acid-alkaline balance in your system. This is a "body clock" rhythm that is needed to metabolize fats, carbohydrates and calories to enable them to become assimilated in the body, rather than stored as weight! The hydrochloric acid enzyme also frees iron from foods and transforms it into a substance that can build good blood and body health. Hydrochloric acid may well be one of your body's most important weight-melting enzyme.

How Overweight Accountant Becomes Slim-Trim In 30 Days

Glenn P. loved to eat meats, beef stews, casseroles, thick roast beef sandwiches, even heavier roasts for weekends. He had developed a paunch. He was starting to sag. He had to keep letting out his belt. He did not walk, but waddled! When he weighed himself, he realized he had 42 extra pounds on his body. But Glenn P. loved hearty meals. He had tried hypnosis, group therapy, starvation diets, various formulas, but they made him all the hungrier. When he went off those diets, he gobbled down so much food, he weighed more than when he had begun those diets! He had to do something. He was getting so heavy, he could hardly get up from his seat. When he either sat or walked, his heavy paunch protruded so that cruel jokesters asked if he were pregnant! He could not endure the taunts. He wanted to go on a program that would let him eat and slim down. A "forever slim" client of his told him how the Enzyme-Catalyst Diet could let him indulge in most of his favorite foods and slim him down. Here is the way Glenn P. followed the ECD Program:

1. *Begin* each meal with a large raw, fresh fruit salad. This is important since raw fruits are a prime source of enzymes. He was told to stuff himself with as many raw fruits as possible. Not only did this help put a stopgap on his runaway appetite, but this gave his digestive system a shower of enzymes that could then attack eaten food.

2. Eat most of your favorite foods, but skim off the gravy, and trim off the fat. *Chew* the stews, casseroles, meat pies, very thoroughly before swallowing. **Benefit:** Chewing will stimulate an even greater flow of vital hydrochloric acid needed to break down the fats of these meats and cause them to become assimilated, rather than built up as fat!

3. Do not drink liquids *with* any meals. Liquids will dilute enzyme power. This paunchy accountant, Glenn P. was told to let his hydrochloric accid attack eaten meats and fats, without liquid interference. This gives "full power" to digestive enzymes to catalyze weight-building fats, carbohydrates and calories.

4. To quench thirst, *sip* a glass of pineapple juice. The benefit here is that pineapple is a prime source of *bromelain,* an enzyme that can attack the toughest kinds of meats and prepare them for better assimilation. But sip a little bit. Do not "drown" your enzymes with liquids. Keep them healthy by protecting them against excessive liquids while they are digesting your foods.

5. *Finish* your meal with another raw fruit salad. Sprinkle with raw wheat germ for more enzyme powers.

Heavy Paunch Melts . . . Fat Is Washed Out Within Four Weeks

Glenn P. could enjoy his meat stews and casseroles on the Enzyme-Catalyst Diet Program. The enzymes actually "melted" his heavy paunch. Within four weeks, or just 30 days, he lost close to 40 pounds! All this because he let enzymes do his reducing for him! Now he is called "Slim" instead of "Tubby" and life is youthfully healthy, once again.

HOW RAW FOODS CAN "SHRINK" FAT CELLS

An assortment of easily available raw foods, found at any local market as well as health stores, can attack the *source* of fat buildup. Namely, in your body's fat cells. In particular raw foods contain enzymes that stimulate and alert body enzymes in your intestinal and glandular networks to instantly "seize" fats, carbohydrates and calories and "work them over" so they are catalyzed and used to build health, rather than build weight.

Bile is an intestinal enzyme that catalyzes fat from milk and cheese and transforms it for better body assimilation, rather than excessive storage in your fat cells.

Lipase is a pancreatic enzyme which metabolizes fats into more usable fatty acids and sends them for better absorption throughout your body to nourish your skin, mucous membranes, nerve linings, brain tissues. Some is stored in your fat cells but not to an excess—provided raw foods stimulate the flow of the lipase enzyme from your pancreas (soft glandular tissue across the back of your abdomen.)

Tripsin is another pancreatic enzyme which works on milk and cheese fats in your intestine to send them throughout your body for better health. It acts as a sentinel in that it guards against fat cell buildup. Raw foods are needed to stimulate the flow of fat-melting tripsin.

Amylase is a digestive enzyme that metabolizes the starches of grains and most bread foods, as well as spaghetti and macaroni. Many overweights who fortify their bodies with the ECD Program and lots of raw foods *before* and *after* a favorite spaghetti, macaroni meal, will be less inclined to fat buildup provided these enzymes are able to attack carbohydrates and protect fat cells from becoming stuffed. Raw foods offer these needed starch and carbohydrate melting enzymes.

Lactase is an intestinal enzyme that breaks down milk sugar, or lactose, in dairy products.

Streapsin is an intestinal enzyme that catalyzes fats from dairy products.

Amylopsin is an intestinal enzyme that catalyzes starches from grain foods and uses extracts for building body health.

Nature has created enzymes for the purpose of transforming foods into a form that can be utilized by the body. An ample amount of enzymes can take heavy and light foods, too, and catalyze them to build your body from head to toe. Enzymes alert your body chemistry to take ingested food and metabolize it, protecting against excessive storage in your fat cells, the basic cause of overweight.

The ECD Program promotes this metabolic action to help your body dispose of excess fat and body fluids, slimming down your waist, abdomen, thighs, legs so that you can lose 10-20-40 pounds while you continue to indulge in your favorite foods.

HOW 36 POUNDS WERE LOST "ALMOST OVERNIGHT" ON THE ECD PROGRAM

Susan A. loved good foods . . . and lots of it. She had a craving for "feast" like foods such as lasagna, meaty roasts, thick goulashes, lamb stews. Naturally, she had a problem with her weight. The scales told her she was close to 50 pounds overweight. More unsightly, her chin sagged, she had a very heavy stomach that a girdle could not conceal, thick thighs that shook like jelly. She had a closet filled with beautiful clothes but could not wear them. She "outgrew" them because she gained weight almost daily. When she began to get dizzy spells, she knew she had to do something. But she loved these foods. How could she slim down? A neighborhood diet specialist told her that a raw food program combined with a slight adjustment in eating practices could help her lose 20 or 30 or even 50 pounds "almost overnight" because the enzymes work fast and steadily. Susan A. might have resisted but just as she loved good food, she loved good clothes and yearned to wear her "too small" fashions in her closet. So she decided to follow the outlined *Enzyme-Catalyst Diet* Program:

1. All food was to be chewed thoroughly. Susan A. habitually bolted down her food and this inhibited enzyme action. Slow chewing enabled her to eat and enjoy food and let her body enzymes pour forth to attack the heavy fats and starches she would soon be devouring. (Slow chewing also put a natural

control on her appetite. She was satisfied with smaller portions, too.)

2. Before beginning any meal, she was to eat a bowl of *raw* vegetables. A raw salad dressing of some polyunsaturated oil with apple cider vinegar and a sprinkle of orange or lemon juice gave her still more enzymes. Chewing these thoroughly alerted her hydrochloric acids to prepare for attacking the heavy meals to come.

3. One day a week, go on a raw food diet. Eat all the *raw* fruits and vegetables you want for that entire day. **Benefit**: Susan's enzymatic system was weak because of her previous lack of raw foods. This raw food plan boosted the secretion of stomach gland enzymes that would strengthen her body's ability to metabolize weight-causing fats, carbohydrates and calories.

4. Dessert was to consist of raw fresh fruit with a sprinkle of honey. **Benefit**: These two foods, raw fruit slices and honey are prime sources of enzymes and require almost *no* body digestion. They are absorbed speedily into the bloodstream, because they are pre-digested. This enables the digestive system to work at metabolizing "heavier" foods, without interference from an otherwise heavy dessert.

5. After each meal, she was told to walk around, calmly, quietly. **Benefit**: Enzymes are better activated if the body is mobile. For very heavy weights, enzymes will need this activity and a few moments of casual walking will alert-acti-vate-amplify their weight-melting powers.

Loses 36 Pounds In No Time At All. This easy-to-follow ECD Program helped her body metabolize fats, carbohydrates and calories, sparing her fat cells from becoming engorged and weighty. She slimmed down so quickly, so easily, it looked as if Susan A. had lost 36 pounds "overnight" or "in no time at all." Her chin was thin, her thighs were slim, her tummy became flat. When she walked, she looked and felt youthful. All this was done without missing breakfasts, luncheons or dinners. No starvation dieting. No strenuous exercise. Rather, she ate broiled meats, lamb chops, meat loaf, liver, omelets and favorite foods...but used the above 5-step "fat melting" program with enzyme boosting action and melted away 36 pounds in no time at all. More important, this ECD Program, followed easily, every day, *kept her youthfully slim—while she ate good food! It was permanent weight loss.*

10 ENZYME STEPS TO
YOUTHFUL SLIMNESS . . . WHILE YOU EAT

Here is a 10-step program, using everyday foods, to feed your body needed enzymes for youthful slimness without denial of favorite foods:

1. *Raw Foods Are Essential.* Plan to eat at least half of your foods in a raw state. This is rather simple. Remember that fruits and vegetables should be raw . . . except those few vegetables which must be cooked. Raw fruits, eaten as snacks as well as desserts, raw vegetables eaten as a platter and also as snacks, can provide your body with a power-house of fat-melting enzymes.

2. *Eat Seasonal Raw Foods.* All raw foods should, preferably, be in season. This is when their enzyme power is at their peak. If you must buy out-of-season foods, select frozen ones. Freezing does not completely destroy enzymes. Rather, it inactivates them. But thawing causes rapid depletion of enzymes so eat thawed foods as quickly after removing from freezer as possible. Otherwise, seasonal raw foods are prime sources of weight-melting enzymes.

3. *Quick-Cooked Vegetables.* Take a tip from the slim Oriental. He cooks vegetables as quickly as possible, in as little water as possible. This preserves enzyme content. Vegetables should be half-raw and crisp for good health.

4. *Steam Vegetables.* Steaming, in which vegetables do not touch water, is an excellent and tasty way to enjoy them cooked. Any housewares store will sell you a "steamer" which you insert in a pot of water just high enough so that it does not reach the vegetables. This lets you cook any vegetable in a steamed manner that gives you succulent juicy good taste with prime enzyme content.

5. *Chew Foods Thoroughly.* Enzymatic digestion and breakdown begins right in your mouth! So take time to chew your foods thoroughly. Strong carbohydrates and fats will be better metabolized and less weighty on your body if you chew thoroughly.

6. *Enzyme Dairy Products.* Wherever possible, select raw certified milk and natural cheeses. Goat's milk is a prime source of enzymes and is available at many health food stores. Avoid processed dairy products since they are devoid of needed enzymes.

7. *Enjoy Tasty Enzyme Foods.* So-called "fermented" enzyme foods include sauerkraut, yogurt, kefir, home-made cheeses. These are prime sources of weight-melting enzymes. They are almost pre-digested and spare your stomach any extra activity. Eat them regularly. They also nourish the bacterial flora in your intestinal-digestive tracts and promote the flow of enzymes for better assimilation.

8. *Raw Grains Are Tasty Enzymes.* Use raw grains. Wheat and seeds are prime sources of enzymes. A variety of any types of raw wheat and seeds used in whole grain cereals and for munching and snacks, can give your body a powerhouse of weight-melting enzymes. These are available at most health food stores.

9. *Use Enzyme Tonics.* Fresh fruit and vegetable juices are prime sources of enzymes. Delicious, too. Bottled and canned juices are usually processed or pasteurized and enzyme content may be reduced. Buy a juice extractor at any health store or department store. Squeeze your own for an adventure in good taste and high enzymes. Drink these raw fruit and vegetable juices daily.

10. *Include Enzyme Foods In Diet.* Health stores have food supplements which are prime sources of enzymes. These include:

• *Papaya.* A fruit that is a top notch source of the enzyme, papain, needed to metabolize heavy protein in meats and main dish foods. Eat the fruit, drink the juice.

• *Brewer's Yeast.* A grain food that is a potent source of B-complex vitamins and plant enzymes that metabolize carbohydrates and fats. Available as tablets or powder.

• *Honey.* Pre-digested, goes to work speedily in boosting enzyme supplies in the system.

• *Rose Hips.* Fruit of the rose after the petals have fallen off. A powerful digestant that works on tough meats and other foods. Sprinkle over fruit dessert.

• *Kelp*. Enzymes from the depths of the ocean. A seaweed plant with a naturally tangy taste. Rich in vitamins and minerals, but also an untapped or little known source of natural enzymes. Sprinkle, as a powder, over most foods.

These food supplements are considered "coenzymes." That is, they boost your body's enzymes in their action upon weight-melting metabolism. Use these foods regularly and your metabolism will be increased so that you should be able to wash out from 10 to 40 or more pounds of unwanted weight. *They work while you sleep, too!*

HOW PETER L. USED ENZYMES
TO RID HIMSELF OF 44 POUNDS

As a telephone lineman, Peter L. had to control his weight. He found it increasingly difficult to go up and down poles. He was losing his agility. He developed a "spare tire" that was unsightly, unhealthy and made him an object of ridicule. If he had to walk more than two blocks, he would wheeze and sputter. All this, at a young age of 49. Peter L. had tried diet pills but they made him dizzy. He went on the other "popular" diets but when he skipped a few days, he put weight back on. He wanted to keep on eating while slimming down his "spare tire." A company official who had lost weight on the Enzyme-Catalyst Diet, told him that he surely could eat heartily and slim down on this simple program:

How to Lose 44 Pounds While Enjoying Foods: Make a slow and gradual transition so that your enzyme-digestive system can gradually be rebuilt. Select a starting date. On that date, gradually eat some raw foods. This is easy to do. Make it a rule that any fruits or vegetables you eat will be raw. Gradually, day after day, keep eating more and more raw foods. Include berries, nuts, grains, seeds. Substitute a raw food for a cooked one, whenever and wherever you can. Beverages should also be raw fruit and vegetable juices. If planning on a heavy beef stew or a favorite goulash, then *begin* your meal with a raw vegetable salad, and *end* it with another raw vegetable salad. This planning also helps to cut down on your eating of heavy foods, but does not deprive you of them, either. It took a specified number of weeks as Peter L. began to lose and lose and lose.

Soon, he took in his belt as his waistline melted. He had lost 44 unwanted pounds. He felt fit, supple, agile. He looked youthful.

Thanks to this simple ECD Program, his "spare tire" vanished and he could enjoy many of his favorite foods . . . while slimming down.

More important to this telephone lineman, the weight he lost, stayed off! By following the rule of eating raw foods daily, he was able to enjoy *permanent weight loss.* His fat cells were slim. So was his body!

HOW TO PROTECT ENZYMES FROM BEING DESTROYED

Enzymes are abundant solely in natural and non-processed foods. Any temperature over 122° F. will kill enzymes. Therefore, if you cook, heat, can, pasteurize or process any food, you automatically destroy all enzymes.

To protect against enzyme destruction, you will need to eat as many foods as possible in their raw and non-processed state. Naturally, you cannot eat a raw steak or egg or fish. These are foods that should be cooked. But a guiding rule is *to eat raw whatever can be eaten raw and cook only what must be cooked.* This rule helps protect you against effects of enzyme destruction. Furthermore, raw foods are very tasty and quite delicious. Raw foods are also much easier to metabolize than cooked foods, so they are easier to digest. This is Nature's plan for giving you body enzymes to build your health and keep you slim.

Important Highlights:

1. Thorough chewing of foods will release a powerhouse of enzymes that can control excessive weight.

2. Raw apples alert the flow of hydrochloric acid to metabolize hard fats and strong carbohydrate foods and protect against fatty build-up.

3. Glenn P. used an easy 5-step ECD Program to lose 42 pounds in 30 days . . . while eating most of his favorite meats, stews, roasts.

4. Susan A. lost 36 pounds "almost overnight" on a 5-step enzyme booster program . . . while enjoying broiled meats, lamb chops, omelets, etc.

5. Follow the easy 10-step program to fortify your body with weight-melting enzymes to keep you slim . . . permanently!

6. Peter L. melted down his "spare tire" on an easy ECD Program that let him indulge in his favorite foods while he followed one simple rule: begin and end meals with a raw

vegetable salad. He lost 44 excess pounds while he enjoyed tasty foods.

7. Nature has put enzymes in raw foods. Avoid unnecessary cooking or processing for abundant weight-melting enyzmes in everyday foods.

4

How Tasty Enzyme Tonics Put a Natural Control on Your Appetite

Overweight is usually the fault of an uncontrollable eating urge. You have an obsessive desire to eat and eat and eat. Whatever satisfaction is experienced after a meal is short-lived. In a little while, you are back in the kitchen, looking for snacks, leftovers or whatever else is available to nibble and munch upon. All too often, a late evening snack turns into a heavy calorie-laden meal. Repeated regularly, this keeps on adding more and more pounds. The excuse is that you are always hungry! This is a mental attitude. Control your "emotional hunger" and you should be able to control your appetite and help take off excess pounds. With the use of raw juice enzyme tonics and elixirs, this can be done. It is the tasty and delicious way to put a natural control on your appetite.

HOW ENZYMES SAY "HALT" TO YOUR RUNAWAY EATING URGE

The urge to eat has its beginning in the brain. Here, there are mechanisms which also tell you to walk, sit, run, lie down. There are other mechanisms that tell you when to eat and also when you have had enough. Deep in the center of your brain is a group of nuclei (cellular components) known as the *hypothalamus*. It is so called because it is at the bottom central portion of another part of your brain, the *thalamus*. The word is derived from the Greek *hypo,* meaning "beneath" or "below" and *thalamus,* meaning "chamber." In this area are controls for the basic biologic essentials as temperature regulation, physical needs as well as the desire to eat. The *hypothalamus,* being composed of millions of cells, also needs a regular and steady supply of enzymes to maintain good health and normal function. When the *hypothalamus* is adequately nourished with sufficient enzymes, it is under better self-control. *Enzymes will promote a food-regulating and satiety reaction upon the hypothalamus and thereby control a runaway appetite.*

How Enzymes Soothe Your Eating Urge. Each brain half has two eating-control centers. One triggers off your eating or hunger urge. The second gives you a feeling of satiety. These eating-control centers are called the *appestat*. Composed of cells, the *appestat* is the trigger that sets off the eating urge . . . and also should later turn off your appetite. These cells "cry out" for nourishment. When fed, they should be satisfied. But if these *appestat* cells are deficient in enzyme nourishment, they cannot be controlled. They keep "crying out" for enzymes which triggers off an eating binge. No matter how much you eat, your cell-starved *appestat* is still "crying out" for more food. The *appestat* wants enzyme satisfaction as a means of limiting the eating urge.

Hunger Is In Appestat, Not Stomach. The satiety centers of your *appestat* control the hunger-eating urge. The feeling of an empty or a full stomach is only a mechanical feeling. It is not the true cause of hunger or essential appetite.

Enzymes in raw foods, especially juices, offer a time-release mechanism by which the *appestat* is fed a supply of glucose (the sugar form to which your body converts most of its fuel) as well as fats, oxygen, vitamins, minerals and complex amino acids. Enzymes act as messengers to transport these substances to the billions of brain cells in a steady rhythm and in regulated amounts so that hunger is controlled. A deficiency of enzymes means that the *appestat* goes hungry and there is an urge for eating that cannot be controlled. Give your *appestat* raw juice enzymes and you can speedily control the desire to indulge in compulsive eating. Enzymes say "halt" to your *appestat* and you can help control your weight.

RAW ENZYME TONICS ARE NATURAL WEIGHT CONTROLS

Freshly prepared raw fruit and vegetable juices send a treasure of enzymes to your billions of body and brain cells to soothe the *hypothalamus* and *appestat* cells. Some benefits include:

1. *Creating Energy.* Enzymes nourish the brain cells, repair damaged cell nuclei through their natural energy or vibration, wherein they cause a reaction without themselves changing in the process. Healthy and repaired brain cells can function better in controlling appetite.

2. *Little Digestion Required.* Raw fruit or vegetable juices require little or no digestion in the stomach. They are absorbed directly into the bloodstream, utilized by the weakest of digestive systems, and sent almost immediately to the brain cells to create speedy appetite control.

3. *Controlling Appetite Within Fifteen Minutes.* In order to digest and assimilate the nourishment in raw foods, it may take several hours. But raw juices work very speedily. Within 15 minutes after drinking a glass of raw fruit or vegetable juice, it is speedily digested, metabolized, absorbed by your bloodstream. Before the 15 minutes are over, the raw enzymes from the juice have been sent speeding to your *hypothalamus* where they deposit glucose, fats, oxygen and other essentials to satisfy your *appestat* and control your runaway appetite.

4. *Alerting Metabolism, Cleansing Internal Organs.* In the over-weight, because of so much excessive food intake, digestive power is often too weak to break down the strong cellular structure of solid raw fruits and vegetables to get at the enzymes. A weighty person may be able to use only a partial digestive power for minimal enzyme removal from solids. But raw juices contain ready-to-use enzymes requiring almost no digestive action. These enzymes help to alert a sluggish body metabolism so that more food is burned and more calories used up. The enzymes then create a combustion of fat, to help neutralize the toxic effects of uric acid and help cleanse internal organs. The enzymes then create a healthful chain reaction between a nourished *appestat* and improved digestive reaction. This "link" helps control the appetite and keep weight from accumulating. It also helps metabolize excessive weight.

5. *Raw Juices Are Powerhouses Vs. Dead Juices Which Are Useless.* Enzymes thrive and flourish with power and vigor in raw juices. When fruits, vegetables, grains are subjected to cooking, any temperature over 122°F. will destroy the enzymes. For example, in order to cook them most vegetables are placed in boiling water,

heated to 212°F. This heat will destroy enzymes. Therefore, raw juices are a "must" for powerful brain-satisfying enzymes. They offer a tasty and delicious way to control your appetite and keep you slim.

THE "SLIM TONIC" THAT HELPS CONTROL CONSTANT HUNGER

Blanche C. was always munching. In addition to her three heavy meals per day, she would indulge in readily available sweets, confections, pastries, little sandwiches, various leftovers, crackers, cookies. Blanche C. was 45 pounds overweight. She did not walk but she waddled! Everything shook. She always complained that she could not control her appetite. The key to her overweight was in a malnourished *appestat*. While she ate plenty of raw fruits and vegetables, her overladen and overburdened digestive system could not accommodate the enzymes. She received partial enzyme nourishment.

Three Vegetables In Juice Form Control Hunger. When she had dizzy spells, she knew she had to do something. Diet pills made her feel worse. Exercise was time-consuming, painful and would take too long to show results. Besides, exercise only made her feel all the more hungrier. Blanche C. admitted that the hunger was largely in her mind. Therefore, she agreed to a special "Slim Tonic" that contained equal portions of three plant foods: carrots, beets and the coconut.

Simple Program: Using a home juice extractor, she would make equal portions of carrot juice, beet juice and coconut juice. She would stir the juices together in an average 8-ounce glass, and drink it *before* each of her three daily meals. It was that simple.

How "Slim Tonic" Controlled Hunger. Enzymes in this combination took up the alkaline elements such as potassium, sodium, magnesium, calcium and iron and used them to soothe the "fat cells" throughout the system. More important, enzymes transported these nutrients to the nuclei of the brain cells, depositing them in the *appestat.* Therefore, the cells of the *appestat* were well-nourished and eased their crying hunger. The urge to eat was lessened.

Drink "Slim Tonic" Before Each Meal. Blanche C. would drink one glass of this "Slim Tonic" before each meal. This helped her to cut down on her portions of food. It was the natural appetite controller. Soon, her meals were cut in half and Blanche C. felt just as satisfied as with double portions! Pounds started to melt away from her heavy body.

Drink "Slim Tonic" Instead Of A Snack. Whenever Blanche C. felt the urge to snack or munch on cookies or crackers, she would drink a glass of this "Slim Tonic." Soon, she found that she was no longer such a compulsive eater. While she would enjoy her food, she would enjoy smaller portions of her favorite stews, casseroles, roasts, even her desserts. She no longer craved cookies or candies. Her *appestat* was satisfied and this created a stress-shield that protected her from the urge to snack all the time.

Loses 50 Pounds Within 45 Days. This "Slim Tonic" so controlled her eating urge, she began to shed pound after pound until she weighed herself to discover she had lost 50 unsightly pounds in less than 45 days. Thanks to raw plant juices, enzymes went to the *cause* of her eating urge. Namely, a starved *appestat* in crying need of enzyme nourishment. Now she is more shapely. She walks like a film star. She is lovely to look at. She feels lovely, too!

HOW TO USE FRUIT JUICES TO CONTROL YOUR APPETITE

Fresh juices made from such citrus fruits as oranges, lemons, grapefruits, tangerines are prime sources of cell-feeding enzymes. When you drink fresh citrus fruit juices, abundant enzymes take up the supply of calcium, phosphorous, potassium, Vitamins A and C, and use these as cellular components to build and rebuild the fragile cells of the *appestat* and the *hypothalamus.* Once the enzymes have used these building blocks for cellular repair, there is a "brain satisfaction" that helps promote a natural appetite control.

Mid-Morning Juice. To stave off the pound-adding "midmorning" snack, drink a glass of equal portions of orange and grapefruit juice. Flavor with a bit of honey, if preferred.

Luncheon Slimmer. Cut down on the urge to over-eat of heavy, fatty luncheons with a glass of fresh grapefruit and tangerine juice *before* you start eating. A half teaspoon of honey is satisfying, too.

Dinnertime Appetite Controller. Because your main meal is usually the heaviest, you need a *strong* enzyme tonic. Mix one glass of pure grapefruit juice with a few tablespoons of tangerine juice. Flavor with one half teaspoon of honey. Stir rapidly. Drink *slowly,* about thirty minutes *before* you start your dinner. The secret here is to give the strong enzymes from the grapefruit a good start in satisfying your *appestat* so that you will naturally want your portions cut down.

ENZYMES + VITAMINS = WEIGHT LOSS

A glass of fresh fruit or vegetable juice is a prime source of enzymes, together with vitamins that work together to help wash out the excess fat from your body cells.

Enzymes + Vitamin A = Digestive Satisfaction. Enzymes in vegetable juices will take the Vitamin A and use it for nourishing the digestive and intestinal tracts so there is a feeling of hunger satiation.

Tasty ECD Tonic: Combine lettuce and carrot juice in equal amounts. Drink regularly throughout the day. Helps perk up the digestive system and put a stop gap upon an uncontrollable appetite.

Enzymes + Vitamin B-Complex = Nerve Satisfaction. Enzymes in plant juices will take the Vitamin B-complex and "coat" the nervous system to insulate you against the urge for nibbling. Enzymes will use the B-complex vitamins to promote better metabolism of protein, fat and carbohydrates to guard against cellular overloading. The enzymes also use the B-complex vitamins in plant juices to catalyze the production of or the function of hydrochloric acid in your gastric juices to promote a satiety feeling that controls the appetite.

Tasty ECD Tonic: In a blender, whiz assorted nuts and seeds, add to any favorite freshly prepared fruit or vegetable juice. Add a half teaspoon of Brewer's Yeast (available at most health stores), stir rapidly and drink whenever you have the urge to eat. Within 30 minutes, the enzymes are working with the vitamins to ease your "nervous urge" and help control your weight.

Enzymes + Vitamin C = Cellular Buildup. Enzymes in a raw fruit drink will take Vitamin C and use it for creating collagen, a glue-like substance which binds together cells throughout your body, especially in your brain, where your *appestat* and *hypothalamus* are situated. Once these "hunger sites" are cellularly nourished, they can resist the temptation to create artificial or unnecessary eating.

Tasty ECD Tonic: Combine equal amounts of any citrus fruit such as orange juice, grapefruit juice, tangerine juice, a sprinkle of lemon juice. Add a bit of honey for a sweet taste. Drink whenever you feel "low" or so-called "depressed" because of overeating problems. Helps ease hunger urge while it perks up your spirits.

Enzymes + Vitamin D = Stronger Bone Cells. Enzymes in a plant juice will promote the utilization of Vitamin D to take up calcium and phosphorus from your intestinal tract and transport these minerals via the bloodstream for deposit in the bones and teeth and the intercellular spongy networks of these body parts. Enzymes build up the internal structure so there is resistance to overeating. The body is thus better able to withstand the change as weight is lost.

Tasty ECD Tonic: Mix two tablespoons of any fish liver oil (halibut liver oil or cod liver oil) together with a glass of mixed vegetable juices. Stir rapidly. Drink slowly. Enzymes will take the Vitamin D from the fish oils and use it to rebuild skeletal components so there is ease in losing weight. (You may use milk, containing Vitamin D, as an oil replacement, if preferred.)

Enzymes + Vitamin E = Fat Barrier. Enzymes in a grain juice will take Vitamin E, combine it with oxygen to create a barrier against fat buildup in billions of body cells. Enzymes also use Vitamin E to strengthen the body cells and insulate them against invasion of pound-causing fats, carbohydrates and calories.

Tasty ECD Tonic: Mix three tablespoons of wheat germ oil with assorted raw vegetable juices. Add a little honey, if desired. This helps send a "shower" of fat-melting enzymes that control fat buildup and also give you a feeling of appetite satisfaction.

Enzymes in raw plant juices work with vitamins to help promote growth, good blood coagulation, formation of intercellular materials, formation of antibodies. Enzymes work with vitamins to promote better food metabolism. They help your body utilize and absorb most other nutrients. This creates a feeling of internal harmony and appetite control. Raw juices are prime sources of vitamins as well as enzymes which work together to help control your appetite and melt away pounds.

HOW RUTH M. USED ENZYME TONICS FOR HER COMPULSIVE EATING URGE

Ruth M. was a self-admitted "compulsive eater." Whenever something went wrong, if her husband told her about business worries, if her children had school problems, if her relatives annoyed her, and even when she had nothing to do, she would start to gobble up as much food as she could put her hands (and mouth) upon. It was a symptom traced to nervous tension.

When her husband started admiring very slim women, when her children expressed embarrassment over hearing their mother called "Fatty," Ruth M. decided she had to do something. Her family life and happiness were being threatened by her compulsive eating.

Satisfies "Sweet Tooth" With Enzyme Cocktail. She began by looking for a natural way to satisfy her "sweet tooth." She did this with a delicious Enzyme Cocktail that was tastefully sweet but added few calories. The enzymes also metabolized carbohydrates and sent them to her *appestat* to "close off" the urge to eat sweets.

"Sweet Tooth" Enzyme Cocktail: 1/2 cup of pineapple juice, 1/4 cup of apple juice, 1/4 cup of orange juice. Add 1 tablespoon of grated apple. Stir vigorously. Drink slowly whenever there is an urge to partake of a sweet such as cake, candy, pie, etc.

Benefits: Enzymes in the juices take up the plant carbohydrates together with the vitamins and minerals and transport them to the billions of body cells, especially the brain cells. Here, the "thirsty" brain cells are satisfied with the sweet taste of this natural juice and then "turns off" the urge to want to eat a calorie-laden sweet.

Satisfies "Fat Taste" With "Low-Fat, High-Taste Tonic." Whenever she was nervous, Ruth M. had a yen for something with a fatty taste. The reason here is that her body cells were strained, her nervous system was frayed and needed a protective covering. This was responsible, largely, for her urge to eat when nervous. She then prepared a tasty tonic that gave her all the tasteful satisfaction of fat but with far fewer calories.

"Low-Fat, High-Taste Tonic." In a glass of fresh vegetable juice, stir four tablespoons of any polyunsaturated vegetable oil. Good sources include corn oil, safflower seed oil, wheat germ oil, peanut oil. (Available at almost any grocery store or health food shop.) Stir vigorously. Drink whenever feeling nervous or the urge to have a taste of fat.

Benefits: Enzymes in the vegetable juice combine with minerals and then take up the polyunsaturated fats from the oil and use these to coat the nervous system; these same fats are transported by enzymes to the billions of body cells and particularly those of the *appestat* to ease the gnawing urge for eating a fatty food.

Loses 38 Pounds While Eating Favorite Foods.

Her compulsive eating stemmed from her overworked nervous system. This triggered off her *appestat* to make a demand for sugars and fats. Enzymes in the various tonics took the natural plant sugars and fats to create cellular repair of the brain cells. This helped create better satisfaction. The gnawing or compulsive eating urge was eased and controlled.

Ruth M. would drink raw plant juices throughout the day. She found that she soon could eat regular size portions with such satisfaction that her excess weight began to melt and melt and melt.

When she shed 38 pounds, her husband felt proud of her. Her

children admired her new and youthful look. Ruth M. was slim and healthy. Her nerves were soothed. No longer was she bothered by the stresses of daily living. The raw juice enzymes had rebuilt her body so she had greater strength of resistance. Now she could cope with stresses . . . and a smile. Enzymes gave her a new lease on a slim life ahead!

HOW TO ENJOY TASTY ENZYME BENEFITS
FROM RAW JUICES

Enzymes in raw juices create a "magic elixir" action in rebuilding body health. These enzymes are able to speedily rebuild damaged and broken body cells, create a barrier to avoid penetration by excess fats, carbohydrates and calories. When prepared in juice form, from fresh fruits and vegetables and grains, enzymes are very active and work speedily.

Avoid Delays. Enzymes may evaporate if the raw juice is allowed to stand for hours and hours after preparation. So avoid delays. As soon after squeezing, drink your tasty enzyme tonics.

Juice Extractor Or Hand Extractor. You may want to obtain an electric juice extractor, available in any department store or health store. It is a simple machine to operate. Insert plant foods, flip the switch, and in seconds, you have a glass of pulp-free juice. Or, you may want a hand extractor. You need to manually work the fruit into the hand extractor and some of the pulp may be in the juice which is flavorful, if you do not mind these particles. Either method will give you delicious enzyme tonics.

Buying Your Foods. Fruits and vegetables should be in season. Wash them thoroughly. Cut out all wilted or rust spots. You may use a stiff vegetable brush to "scrub" away any stains. Slice the plant foods into small pieces and then insert in your extractor.

Sip, Taste, Enjoy. Hold a glass of fresh juice up to your lips. Focus thoughts upon the juice. This alerts your digestive enzymes to get ready for the beverage. Then slowly sip, and taste and swallow the juice. This helps condition your digestive system and prepare it for transporting the enzymes to your billions of body cells. Treasure each sip.

Drink Fresh Juices. Exposure to air, light and oxygen may reduce the enzyme and nutrient power of the juices. You should plan on drinking them as quickly after squeezing as possible.

If You Must Store Juices. Place the juice in a glass jar. Cap the jar with a tight-fitting lid and keep on the top shelf of your refrigerator where you would keep milk and other beverages.

Room Temperature Is Healthier. When removing chilled juices, let remain at room temperature for a little while. An icy beverage *constricts* digestive organs and inhibits the free transport and flow of enzymes. Let very cold beverages stand at room temperature until comfortably cool. Your digestive system will be more receptive to *comfortable* beverages that are neither too icy or too hot.

By including healthful enzyme tonics of fresh fruit and vegetable juices, you can help soothe your hunger pangs and ease your compulsive eating urges. Enzymes offer satisfaction to your *appestat* and help you control your appetite, the natural and tasty way!

Special Highlights:

1. Enzymes promote a food-regulating reaction upon your *hypothalamus* and *appestat* in your brain.

2. Raw enzyme tonics offer five different methods of appetite control.

3. Blanche C. used three juices to control her constant hunger. The "Slim Tonic" helped her lose 50 pounds within 45 days.

4. Use fruit juices to control your appetite for each of your three daily meals.

5. Enzymes + vitamins = weight loss, with fresh fruit and vegetable juices. Make your own various ECD Tonics for tasty weight loss.

6. Ruth M. used enzyme tonics for her "sweet tooth" and "fat taste" urges. She slimmed down while she ate favorite sweet and fatty foods. She lost 38 pounds on this simple program.

7. Enjoy tasty enzyme benefits from raw juices easily prepared in your own home. It's the natural and delicious way to control your appetite and lose excessive pounds.

5

How Everyday ECD Foods
Help Melt Away Bulky Fat

Overweight is often the result of stored up bulky fat in the body cells that is in need of enzymatic metabolism. When you eat foods containing fat (such as meat, fish, eggs, dairy products, desserts), it is transported from your digestive system to the waiting fat cells throughout your body. The fat then accumulates and gathers in tiny reserve droplets within your individual cells. These cells act as living silos. That is, they are *storage depots* for fats. Some fats will be used to provide energy and transport vitamins. But an excess amount of fats or the absence of sufficient enzymes will cause more and more to be stored in the cells. Weight buildup continues. You step on the scale and it keeps going higher and higher.

Stored Fats Need Enzyme Melting For Weight Loss. To help melt away bulky fat, enzymes from raw foods are needed in daily eating. These enzymes penetrate through the cell wall, enter into the gray cytoplasm, within the nucleus where the fat is stored. These enzymes create a *penetrating action* to help dissolve the clumps and, literally, wash them right out of the cells. The raw food enzymes bathe the adipose tissues, penetrate the greatest density of the adipose-storage (fat) cells, break up the accumulated masses and start the process of catalyzing them so that the cells will become slimmer and the process of weight loss has begun. This Enzyme Catalyst method works with lightning swift precision. Moments after you have taken an enzyme food, these catalysts are at work in helping to free your billions of fat cells from accumulated pound-causing weight.

HOW A 5¢ ENZYME FOOD HELPS MELT AWAY BULKY FAT

John O. was a typical "meat and potatoes" man. As a factory worker, he was raised on heavy meals. He could not easily give up this craving for thick stews, meaty casseroles, juicy roasts, thick steaks and the like. It was rare that John O. could do without such heavy meals. Even lunchtime called for a goulash or a plate of heavy meats with boiled potatoes and cabbage. Years of such eating had put on a "spare tire" around his waist. The company doctor told him that he was 44 pounds overweight. But John so loved his heavy foods, he could not cut them out. But when he saw how slim his factory foreman kept, even though he, too, ate the same foods, he asked for the secret.

The factory foreman and John O. ate the same foods daily, but the foreman was slim, while John O. was 44 pounds overweight and still climbing. The foreman's secret? An everyday food, available in any market for a small cost. Just two tablespoons daily cost about 5¢. It kept the foreman slim while he ate heartily. *It was apple cider vinegar.*

Simple Program Melts Pounds. The secret was to take two tablespoons of apple cider vinegar in a glass of any beverage such as a raw fruit or vegetable juice *before* a meal. Or, just sprinkle it over a raw salad. Or, add two tablespoons of apple cider vinegar to two tablespoons of any polyunsaturated oil and sip slowly *before* a meal. These two ingredients can also be added to a fruit or vegetable juice as a before-meal enzyme cocktail.

Benefits Of 5¢ Enzyme Fat-Melting Food: Apple cider vinegar is a fermented food. It is a concentrated and powerful source of enzymes which burn the accumulated fat in the adipose tissues so there is little excessive storage. Enzymes in the apple cider vinegar favors oxidation of the bloodstream, maintains better mineral metabolism, promotes a healthful fat-dissolving action that lets you enjoy your favorite foods without fattening up.

Appetite Controlled, Paunch Melts Down. John O. followed this simple program. Every day, before each of his three meals, he would take two tablespoons of apple cider vinegar in a raw fruit or vegetable juice. He found that his appetite was controlled. But more important, even if he ate his luscious roasts or satisfying stews, the heavy feeling was dissipated, and his waistline started to melt, so to speak. Within six weeks, John O. had melted down 50 pounds. His paunch was gone and now he had to buy new clothes for his slim-trim figure. He still enjoys his delicious "meat and potatoes" fare but uses the simple *5¢ Enzyme Fat-Melting Food,* the apple cider vinegar, before each meal as a natural appetite controller and also fat-melter!

HOW FERMENTED FOODS ARE ENZYME CATALYSTS

Apple cider vinegar is a *naturally fermented* food, available in almost any supermarket or health store. Select the bottle that is marked *apple cider vinegar.* Do NOT use ordinary vinegar or any distilled varieties which have the living enzymes destroyed.

Enzyme Catalyst Power: Fermented apple cider vinegar contains enzymes which create a catalyst action on body fats. Enzymes in fermented apple cider vinegar helps attack (and when used with polyunsaturated oil) dissolves larger quantities of accumulated fat. The decomposition of fat in the adipose cells is catalyzed by the same chemical process that tenderizes meat soaked in vinegar, or marinated in oil and vinegar. Enzymes in the apple cider vinegar act as a solvent of fatty substances. When you drink the apple cider vinegar, even in such small quantities as two tablespoons in a raw juice, its enzymes create an emulsifying action on accumulated fat in the cell tissues.

Dissolves Solid Fat. Fermented apple cider vinegar has extra-powerful enzymes which act as a lipotropic agent or fat disperser. The powerful fat-dissolving action of this 5¢ enzyme food is mentioned (although not so scientifically) by Pliny, the early Roman author and naturalist. He told how Cleopatra won a bet when she dissolved pearls in vinegar. Since pearls are lipidlike (fatty) tumescent exudations that have solidified, layer by layer, it appears obvious that the enzymes can dissolve solid fat in the body. Since vinegar has the natural ability of dissolving the solidified fatty layers of a pearl, it can also help "dissolve" or "wash away" the much softer but solid layers of human flab and fat. It is an ancient secret that is a newly discovered method of using enyzmes to keep slim.

HOW TO MAKE YOUR OWN ENZYME FOOD

You can enjoy home-made apple cider vinegar. To make, put peelings, cores and sliced apples in a wide-mouthed crock. Fill with cold water to cover. Now cover the crock with a lid. Keep stored in a warm place. Add fresh peelings, cores and sliced apples from time to time. You can "see" the enzymes in the form of a slowly thickening substance that accumulates on the top. This is the prime source of fermented enzymes, containing the bacteria of the genus *Acetobacter.* When the vinegar in the crock tastes strong enough (after eight to ten weeks), strain it, bottle and cork it and use regularly. The viscous, gelatinous substance that forms on the top may be removed and used as a starter in speeding the action of the vinegar in another batch.

HOW SAUERKRAUT CAN HELP MELT AWAY BULKY FAT

As a fermented food, sauerkraut has its beginnings as cabbage, a prime source of vitamins, minerals and fat-melting enzymes. When fermented into sauerkraut, the process helps develop desirable bactericides (natural antibiotics) and stronger enzymes which are used by the body for better health and also for cleansing the adipose tissues in which fat is stored. As a fermented cabbage food, sauerkraut can help promote the enzymatic penetration into the adipose tissues and control weight.

How To Make High-Enzyme Fermented Sauerkraut: Remove outer green leaves of cabbage. Use for soups and stews. Remove the core. Shred the cabbage or cut into thin slices. Place into a glass jar or clean crock. Add sea salt (sold at health stores) or any salt substitute in the amounts of one pound to forty pounds of cabbage. Add water to cover. Then cover with a plate or clean cheesecloth. Store in a cool place. Soon, the salt will start taking out a lot of cabbage juice from the cabbage. As the cabbage ferments into sauerkraut, the process causes development of powerful enzymes. Later, when you take out the sliced or shredded cabbage, you will have fermented sauerkraut. Use it regularly as part of your raw salad. The enzymes are potent and have the ability to create a catalyst action in the body so that fat can be dissolved within the adipose tissues and energize the *mitochondria,* or power centers of the cell. These power centers are activated by the sauerkraut enzymes and help melt away excess or accumulated fatty deposits.

HOW TO EAT ALL YOU WANT OF
(MOST) FATTY FOODS AND KEEP SLIM

The Enzyme-Catalyst Diet Program calls for fortifying your system with a regular supply of these fat-melting ingredients, so that you can protect yourself against weight build-up. Let us take a closer look at fats in your foods and how enzymes will enable you to eat all you want of (most) of your favorite fatty foods and still keep slim.

You Need Fat. Yes, fat is important for your health. Fat is one of seven groups of necessary nutrients required for life and health, (the other six are vitamins, minerals, proteins, carbohydrates, water and enzymes.) Fat helps give you energy, padding, conserves body heat, transports vitamins, offers a satiety value in meals, acts as an intestinal lubricant, builds body cells, soothes nerves, provides valuable essential fatty acids and offers food flavor. You do need fat. But you also need enzymes to metabolize the fats. You will also need to make a few adjustments in your eating program to build health, control weight and enjoy most of your favorite fats in daily eating.

The following charts tell you which foods are high in saturated (hard) and unsaturated (soft) fats:

A Word of Explanation about Fats

There are three major types of fats: saturated, polyunsaturated and monounsaturated. We need them all, but in proper amounts. *Saturated fats* are found mainly in meats and dairy products; *polyunsaturated fats* are found in fish and most vegetable oils (corn, soybean, peanut, cottonseed and safflower oils, but *not* coconut oil which is high in saturated fat); *monounsaturated fats* are found in poultry (chicken, duck, turkey) and olive oil.

If saturated fats are eaten predominantly, the level of cholesterol in the blood may increase. Moderate use of saturated fats plus increased use of the polyunsaturated ones tend to lower the cholesterol level in the blood. Polyunsaturated fats also supply linoleic acid, an essential fatty acid which the body cannot make itself. Many foods contain both saturated and unsaturated fats. A partial list* of common foods according to type of predominant fat follows:

Foods According to Types of Fats

Predominantly Saturated Fats	Predominantly Polyunsaturated Fats	Predominantly Monounsaturated Fats
Meat—beef, veal**, lamb, pork and their products such as cold cuts, sausages Eggs Whole milk Whole milk cheese	Liquid vegetable oils*** corn, cottonseed, safflower, soybean Margarines containing substantial amounts of the above oils in liquid form	Olive Oil Olives Avocados Cashew nuts

Cream, sweet and sour
Ice cream
Butter
Some margarines
Lard
Hydrogenated shortenings
Chocolate
Coconut
Coconut oil
Products made from or
    with the above, such
    as most cakes, pastry,
    cookies, gravy, sauces
    and many snack foods
Shellfish

Fish
Mayonnaise, salad
    dressing
Nuts—walnuts, filberts,
    pecans, almonds,
    peanuts
Peanut butter
Products made from or
    with the above

---

*Based on *How To Follow the Prudent Diet,* New York City Department of Health, revised July, 1969.

**Veal and poultry (chicken and turkey) are relatively low in total fat. Veal fat is predominantly saturated; chicken and turkey fat is more favorably distributed between polyunsaturated and saturated fat.

***Peanut oil is not polyunsaturated to the same degree as the other oils.

Foods High in Saturated Fat Content

Meat (beef, veal, lamb,
    pork, sausages)

Egg yolk

Whole milk

Whole milk cheese
    (American, Swiss and
    other hard cheeses)

Cream and cream cheese

Ice cream

Butter

Some margarines

Solid shortenings

Chocolate

Coconut, cocnut oil

Foods prepared with above
    products (most cakes,
    pastries, cookies and
    gravies)

Foods High in Unsaturated Fat Content

Fish

Seafood

Margarine containing
    substantial amounts of
    liquid vegetable oils

Mayonnaise and oily salad
dressing

Liquid vegetable oils (corn,
cottonseed, soybean,
safflower and others)

Walnuts, almonds, pecans,
filberts, peanuts and
peanut butter

Note: Chicken and turkey have about equal amounts of saturated and and unsaturated fats.

Enzymes Help Dissolve Fat. Basically, fat contains the elements of carbon, hydrogen and oxygen in various combinations of animal and vegetable fats. *Fat does not dissolve in water.* It dissolves through the action of enzymes which penetrate the outer surface covering, and brings about a "melting" of its carbon, hydrogen and oxygen. When you eat a raw fruit or vegetable with a fat food, you are giving your body the enzymes needed for dissolving the fats that will be stored in your body, as a result of the fat food.

Enzymes Help Control Cholesterol. Among the fatty substances believed to be casual factors in the development of heart trouble, *cholesterol* is considered to be a major instigator. Cholesterol is a tasteless, odorless, white fatty substance found in all animal fats and foods of an animal source. It does not dissolve in water. Enzymes, however, can help control cholesterol buildup which is also a problem in overweight.

Raw fruits and vegetables are prime sources of those enzymes which are sent to the liver (where cholesterol is synthesized from substances derived from eaten fats, proteins and carbohydrates) where they act to control its manufacture. Enzymes can help dissolve cholesterol in the liver and other body parts and thereby control weight and also protect your heart.

Enzymes Tame Triglycerides. Another coronary risk factor, and contributor to weight is a fatty substance known as a *triglyceride.* This is a substance found in most fatty foods. Also, if an excessive amount of sugar is consumed, it is turned into triglyceride in the body. Sugar may raise blood triglyceride, just as saturated fat tends to raise blood cholesterol. This can also cause serious weight buildup. Triglycerides do not dissolve in water. But raw food enzymes, as well as enzymes from fermented foods, can penetrate the sheath of the fatty triglycerides. These enzymes help dissolve these fats so there is greater protection against weight buildup, coronary disorder and hypertension, too.

YOUR ENZYME-CATALYST DIET PROGRAM
FOR KEEPING SLIM WHILE EATING FATS

The ECD Program calls for a few adjustments that will help you keep slim while eating fats. The clue here is to eat the *right kind of fats* that offer you good taste, good health and are easy for enzyme metabolism.

Begin Each Fat Meal With Super-Active Enzyme Foods. Since heavy fats require stronger enzymatic action, begin your meals with super-active enzyme foods such as: cranberries, grapefruit wedges, orange wedges, salad of lemon, lime and nectarines, pineapple slices, sour grapes, strawberries, tomato slices. Eating such super-active enzyme foods sends a supply of vital enzymes throughout your digestive system to await the fat foods for better digestion.

Basic ECD Program

Eat These:

Very lean, well-trimmed meats:
Beef: Roasts (rump, heel, sirloin tip) and steaks (round, tenderloin, flank ground round or sirloin.)
Veal: All cuts except breast.
Pork: Best to avoid all cuts.
Lamb: Leg or loin roast and chops.
Poultry: All cuts. Remove skin and discard.
Fish: Fresh and salt water fish without skin.
Beans, Peas, Nuts: Kidney, lima, navy and pea beans, lentils, chick peas (garbanzos), split peas, soybeans.
Eggs: Egg whites as desired.

Avoid or Use Sparingly:

Heavy marbled and fatty meats, spare ribs, mutton, frankfurters, sausages, fatty hamburgers, bacon, luncheon meats.
Avoid organ meats such as liver, kidney, sweetbreads. (Since liver is very high in iron and vitamins, it should not be eliminated from the diet entirely. Use a 4-ounce serving in a meat meal no more than once a week.)
Avoid: cakes, batters, sauces and other foods containing egg yolk. Egg yolks: limit to 3 per week, including eggs used in cooking.

About Fats and Oils

Eat These:

Margarines, liquid oil shortenings, salad dressings and mayonnaise containing any of these polyunsaturated vegetable oils: corn oil, safflower oil, sesame seed oil, soybean oil, sunflower seed oil. All margarines should be made with one of

Avoid or Use Sparingly:

Coconut oil, coco butter, palm oil, hydrogenated or "hardened" vegetable shortenings, meat drippings, suet or lard. If you use butter, mayonnaise, meatbased gravies and sauces, do so sparingly.

the above oils. Diet margarines are
low in calories because they are low
in fat. Therefore, it takes twice as
much diet margarine to supply the
polyunsaturates contained in a
regular margarine.

Note: Salad dressings should have no egg yolks and be made with
one or more of the above oils. Homemade sauces and soups should
be made with polyunsaturated fats and skim milk or skimmed stock.

A 10-STEP ENZYME-CATALYST DIET PROGRAM FOR A FAT-LOVING OVERWEIGHT

Gloria T. was fat as a child. She could not endure the rigors of
dieting on so many different programs that called for serious
restriction of fats. She loved to eat fat. She had slimmed down on
some diets, but as soon as she went off the diet, she started to gain
unsightly weight again. She was some 38 pounds overweight. But
particularly unsightly were her protruding rear, her very heavy
bosom, the thick thighs. Also, Gloria T. was a short five feet in
height so she looked even fatter than an ordinary person. But she
loved fat so much, she could not give it up. She always said she was
"born fat."

Baby Fat Into Adult Fat. The total number of adipose (fat) cells is
established in your youth. If you were overfed as a baby, you will
have a larger number of fat cells than normal. That total remains
constant as you grow into adulthood. But when a "baby fat" child
such as Gloria, became an "adult fat" person, she did not increase
the number of her fat cells. She kept filling up the ones she had! On
ordinary diets, when she lost weight, her fat cells would shrink. They
did not disappear. What her fat cells needed was a daily substance
that would wash out the excess fat from the connective tissues and
keep it out! This is where the ECD Program helped get to the *root
cause* of her overweight; namely, the fat cells. But more important,
she could eat enzyme foods daily to keep the fat reduced and remain
permanently slim . . . while eating her favorite fatty foods!

Here is the 10-step Enzyme-Catalyst Diet Program that melted 38
pounds (and more) from Gloria T. while she could enjoy almost all
delicious fatty foods:

1. She could eat a maximum of three egg yolks a week, but
   before eating any food containing egg yolk (or egg, itself),

she had to eat a small peeled grapefruit or two small oranges for powerful fat-digesting enzymes.

2. She had to limit the intake of shellfish and organ meats because of high cholesterol counts. But when she partook of fish, she would "enzyme-enrich" it by marinating it in orange juice. She would also squeeze lemon juice atop fish before eating it. This added flavor and enzyme enrichment to the fish for better metabolism.

3. Gloria T. enjoyed chicken and turkey quite frequently. These would be eaten with fresh pineapple slices for better taste and speedy enzyme digestion.

4. When selecting lean cuts of meat, she would trim away all visible fat. (She also discarded the fat that cooked out of the meat.) She whould eat this meat with fresh apple and pear slices for better enzyme metabolism. This gave her all the luscious taste of the fat she loved, but with less weight-buildup.

5. She avoided deep fat frying. (This coats food with heavy fat and enzymes cannot penetrate too deeply.) Gloria T. enjoyed foods that were baked, boiled, broiled, roasted or stewed. Always, she would eat a raw vegetable salad with a salad dressing made of equal portions of apple cider vinegar and oil. Her enzymatic system was fortified with this natural dressing and fat could be better metabolized and washed out of her adipose tissues.

6. She restricted the use of "fatty luncheon" and "variety meats" such as sausages or salami. Instead, she would have a lean roast beef and tomato platter for good taste and enzyme vigor.

7. Instead of butter and other cooking fats that are solid or completely hydrogenated, she would use liquid vegetable oils and margarines that are rich in polyunsaturated fats. Together with a spoon or two of apple cider vinegar, she could have the taste of fat, but with little weight buildup because of better enzymatic catalyst action.

8. Instead of whole milk and cheeses made from whole milk and cream, she would use skimmed milk, and skimmed milk cheeses. With these, she would also use fresh seasonal fruit which had enzymes that would work to break down

whatever fat was present in the dairy products to prevent against excessive weight buildup.

9. She could indulge in as many cooked egg whites she wanted since it was completely free of cholesterol. She would make scrambled egg whites together with shredded apples for a delicious breakfast that had a good "fat" taste with needed enzymes for fat metabolism.

10. To indulge in the urge to snack, she would avoid crackers that had saturated fats. Instead, she would enjoy bread sticks, rye wafers, melba toast. Chewing these hard crackers alerted her enzymatic system so that these catalysts could be summoned forth to promote better digestion and better fat melting.

Loses 38 Pounds, Enjoys "Fat" Taste In Foods. Gloria T. soon shed those unsightly 38 pounds. She is determined to lose more, while she enjoys the "fat" taste in foods. She is using raw fruit and vegetable enzymes to help wash out the fat in her adipose tissues. Soon, the enzyme program will give her a silhouette figure she always craved . . . and she can continue eating to her satisfaction.

"SOFT" FATS ARE HANDLED EASIER BY ENZYME CATALYSTS

"Soft" or polyunsaturated fats are handled much easier than "hard" or saturated fats by enzyme catalysts. Your food program should use these "soft" fat foods, as listed in the charts, as frequently as possible as substitutes for the "hard" fat foods. You will be satisfying your taste for fat, your need for fat, just the same. But you will also enable your enzyme catalysts to better metabolize the fats and guard against unhealthy weight buildup. It's the tasty way to become slim and stay slim!

Summary:

1. Enzymes create a penetrating action on stored fats and help dissolve them out of the body's adipose (fat) cells and cause natural weight loss.

2. A 5¢ enzyme food helped John O. lose his "spare tire," and some 44 pounds, while he continued to enjoy his favorite "meat and potatoes" type of meal. The Enzyme-Catalyst Diet kept the weight off, too.

3. Fermented foods are potent sources of concentrated fat-melting enzymes.

4. Sauerkraut can help melt away bulky fat.

5. Enzymes enable you to eat most fatty foods and still keep slim. Enzymes help dissolve fat, control cholesterol, tame triglycerides, keep slim.

6. The basic ECD program lets you enjoy most foods while slimming down.

7. Gloria T. lost over 38 unsightly pounds on a 10-step Enzyme-Catalyst Diet Program while still enjoying the taste of fat in her foods.

6

How to Use the ECD Program to Burn Up Unwanted Calories

Barbara U. thought she was a successful dieter. She would indulge in her favorite calorie-high foods, especially sugar, gain up to 30 or even 40 unsightly pounds. Then she would go on a "semi-starvation" diet and lose the excess pounds and become youthfully slim again. But here is where her success ended. As soon as she went off the diet, she gained those pounds back on again. She knew that she was eating far too many calories, but she could not burn them up on any other program except to "semi-starve" herself on minimal amounts of food. She faced a "see-saw" dilemma wherein her weight went up and then went down, because of calories. She wanted a program that would let her enjoy sweets but without calorie buildup.

How Enzyme Catalyst Foods High In Calories. Barbara U. tried a simple ECD Program that called for raw enzyme foods both *before* and *after* eating a high-calorie foods. The program had three basic steps:

1. Before eating any high-calorie food, eat a fresh raw fruit or vegetable. Chew thoroughly before swallowing.
2. Try to reduce the amount of the high-calorie food.
3. After eating any high-calorie food, eat a fresh raw fruit or vegetable. Chew thoroughly before swallowing.

Calorie-Burning Action: Raw food enzymes create what is known as a *high osmotic pressure* within the adipose (fat cell) tissues. This is a cellular penetration action wherein enzymes are able to pierce the outer calorie-engorged layer of the cell, promote a form of osmosis to dissolve the accumulated weight and create a calorie-burning action that will promote weight loss. The osmotic pressure causes the cell to metabolize its calories and guard against build-up. To alert this *high osmotic pressure,* enzymes are needed. These enzymes are found in fresh raw fruits and vegetables and their juices. They are also found in raw grain foods, many seeds and nuts that have *not* been processed. Enzymes, therefore, promote a *high osmotic pressure* on the cells and create a calorie-burning action that does more than bring down weight. *It keeps weight off, permanently, because it reduces the weight of the cells where fat has its source!*

Wins Battle Of Calories With Enzymes. Barbara U. was able to take off at least 40 pounds, while enjoying most of her favorite foods, by following the easy 3-step ECD Program outlined above. Now she is more than youthfully slim. *She is permanently slim,* thanks to the calorie-burning power of enzymes.

HOW ENZYMES KEEP YOUR "CALORIE BANK" IN BALANCE

Think of your body as a bank in which you can deposit or withdraw calories. You deposit calories by eating sugary foods. You withdraw calories by using them up in the form of energy. You keep them withdrawn by including fresh raw foods at each and every meal. This will give your "body bank" a special dividend. The enzymes in the raw foods will burn up calories and your weight will be maintained at a slim level.

Balance Your Calorie Account. The reason for your overweight is as simple as your bank account. If you put in more than you take out, your account grows—and vice versa. If you have stored too many calories and have neglected to deposit raw food enzymes, then you will start to gain weight. Bringing your account down is basically a matter of putting in more enzymes than calorie foods, so that the *high osmotic pressure* can use up the calories and keep your account in balance. You will also lose weight and keep it off if you follow the delicious daily practice of eating raw enzyme foods if and when you must eat calorie foods. This helps balance your calorie account.

BE ALERT TO OVERBANKING YOUR CALORIES

Because adding calories to your account is so easy, you gain weight. Chances are that you are not even aware of many extra calories you are depositing. The reason is "unconscious calories." You come by them through routine methods that are so natural, you are not aware of them. You eat leftovers, you eat between meals, you constantly snack. Calories, no matter how few, have a way of adding up.

If you are to control successfully the calories you eat, you will have to be alert to every calorie that you eat. Excessive calories not only add up, but they throw your calorie/energy account out of balance. Instead of a sugary sweet, reach for a raw food to give you enzymes that boost metabolism and cause calorie withdrawals.

PLANNING YOUR CALORIE-BURNING PROGRAM: OFFICIAL U.S. GOVERNMENT RECOMMENDATIONS

Here is a 5-step calorie-burning program, as recommended by Prof. Katherine H. Fisher of Pennsylvania State University,[1] which is designed to create caloric oxidation, or an enzyme activity:

Excess weight throws out of balance the body's entire system of energy exchanges. For that reason alone, it is a hazard. At least five factors must be considered if one is to lose weight successfully:

First, one should consult a medical doctor to determine whether or not weight reduction is desirable from a health standpoint.

Second, one must want to lose weight.

Third, the amount of weight to be lost and the rate of loss should be decided upon and approved by a medical doctor or a nutritionist.

Fourth, the diet used for weight reduction should be nutritionally adequate in protein, minerals and vitamins as well as calories. (This calls for raw foods which are enzyme sources for oxidation of calories.)

Fifth, after weight is reduced, the new weight should be maintained by controlling the intake of calories and having enough physical activity regularly.

U.S. GOVERNMENT FIGURES ON AMOUNTS OF CALORIES THAT BUILD FAT

Prof. Fisher then says, "Since it takes a deficit of 3500 calories to lose 1 pound of body fat, a daily reduction of 500 calories will result in the loss of 1 pound of body weight each week. 500 (calories) times 7 (days) equals 3500 calories." You can plan on losing 1 pound of body fat for each 3500 calories you burn up! The greater your enzymatic intake, the greater the caloric oxidation reaction will take place and the better your weight loss will be.

---

[1] *Food, Yearbook of Agriculture,* U.S. Department of Agriculture, Washington, D.C.

OFFICIAL U.S. GOVERNMENT LOW-CALORIE, HIGH-ENZYME SLIMMING PROGRAM

Prof. Fisher then offers a sample daily menu that is low in calories (1500 total calories for all three meals) but gives you 40% fat, 35% carbohydrates and 25% protein, and ample raw food enzymes that will help burn up excessive calories:

BREAKFAST: One-half cup of fresh orange juice; 1 cooked egg; 1 slice of bread; 1 teaspoon of butter; 1 cup of skim milk.

NOON MEAL: Four ounce broiled beef pattie; two-thirds cup of green beans with 1 teaspoon of butter; 1 raw apple; 1 cup of skim milk.

DINNER: Four ounce serving of broiled halibut; two-thirds cup of carrots with 1 teaspoon of butter; one-half slice of bread with 1 teaspoon of butter; 2 peach halves with 2 tablespoons of syrup; 1 cup of skim milk.

Tasty Variations: Prof. Fisher says, "Other meats, fruits and vegetables may be substituted in this pattern to give variety and to fit individual food likes and dislikes. A 4-ounce serving of any lean meat, fish or poultry can be used in this diet plan, but the fatty kinds should be avoided. Cheese may replace part of the meat. A medium slice or a 1-inch cube of cheese or 2 tablespoons of cottage cheese constitute a 1-ounce serving."

Avoid These Foods. Prof. Fisher suggests, "One should avoid certain foods in any weight-reduction program—candy, jelly, jams and other sweets; rich desserts (cake, pie, ice cream and pudding); bottled salad dressings (mayonnaise and French dressing); and rich gravies and dessert sauces."

HOW TO KEEP A CALORIE DIARY

So that you can see where these extra and "unconscious calories" come from, here is a three-day Calorie Diary. In it, you list everything you eat each day, at breakfast, lunch, dinner and everything else you eat in between—mid-morning, afternoon and evening.

Be Honest. Do not leave out anything, no matter how small it may be. You must be very honest with yourself if you really want to lose those extra pounds.

Supply And Demand. Weight control follows the rules of supply and demand. You will *lose* weight if you supply your body with fewer calories than its activities demand. (Enzymes will use stored calories to make up the difference and supply you with energy.)

3-Day Food Diary (Before Dieting)

- Be honest with yourself.
- Fill in completely.
- Estimate servings accurately.
- Fill in calories last.

	1st Day		2nd Day		3rd Day	
Food	Calories	Food	Calories	Food	Calories	
Breakfast						
Lunch						
Dinner						
Snacks						
Mid-morning						
Afternoon						
After Dinner						
Total Calories						

Activities: What did you do to "burn up" calories each day and how long did you do it?

You will *keep* your weight at a desirable and permanent level if you have sufficient enzymes available to metabolize excessive amounts of calories.

You will *gain* weight if you supply your body with *more* food calories than enzymes can metabolize, especially if you are deficient in these catalysts. Your body will store the excess calories as fat.

Plan Your Calorie Diary: It is a good idea before you start your Enzyme-Catalyst Diet Program to figure out just how much you are accustomed to eating. Note the 3-Day Food Diary (Before Dieting) chart in this chapter. Put down everything you eat for three consecutive days (including at least one typical weekend day). Then figure out the calorie count as accurately as you can. Use the calorie tables in this chapter. This will tell you how many calories you are eating and how many you have to cut down.

HOW MANY CALORIES DO YOU NEED?

The daily calorie needs shown in the tables are for your guidance. No two people are alike. In addition to height and age, your daily calorie requirement depends upon body frame, present weight, desired weight, occupational activities and other daily activities at home and at play.

Daily Calorie Needs for Men of Normal Weight*

Height	Age 15-19	Age 20-29	Age 30-39	Age 40-49	Age 50-59	Age 60-69	Age 70-79
5ft. 0	2,620	2,250	2,100	2,020	1,980	1,710	1,570
5ft. 1in.	2,690	2,310	2,160	2,070	2,020	1,750	1,610
5ft.2in.	2,750	2,390	2,220	2,110	2,070	1,790	1,650
5ft. 3in.	2,820	2,450	2,280	2,160	2,110	1,830	1,690
5ft. 4in.	2,880	2,500	2,340	2,200	2,160	1,880	1,740
5ft. 5in.	2,940	2,560	2,400	2,260	2,200	1,920	1,780
5ft. 6in.	3,000	2,620	2,460	2,320	2,250	1,950	1,810
5ft. 7in.	3,070	2,680	2,520	2,380	2,310	2,000	1,850
5ft. 8in.	3,140	2,740	2,580	2,440	2,370	2,060	1,900
5ft. 9in.	3,200	2,800	2,640	2,500	2,430	2,100	1,930
5ft. 10in.	3,280	2,880	2,710	2,560	2,490	2,160	1,990
5ft. 11in.	3,360	2,950	2,790	2,620	2,550	2,210	2,040
6ft. 0	3,440	3,030	2,860	2,680	2,610	2,250	2,070
6ft. 1in.	3,520	3,130	2,940	2,740	2,670	2,310	2,130
6ft. 2in.	3,600	3,180	3,010	2,800	2,730	2,370	2,180
6ft. 3in.	3,680	3,250	3,090	2,860	2,790	2,410	2,220

Daily Calorie Needs for Women of Normal Weight*

Height	Age 15-19	Age 20-29	Age 30-39	Age 40-49	Age 50-59	Age 60-69	Age 70-79
4ft. 9in.	2,080	1,890	1,810	1,760	1,710	1,480	1,370
4ft. 10in.	2,110	1,920	1,840	1,790	1,740	1,510	1,400
4ft. 11in.	2,140	1,950	1,870	1,820	1,770	1,530	1,430
5ft. 0	2,190	1,980	1,900	1,850	1,800	1,550	1,450
5ft. 1in.	2,240	2,020	1,940	1,890	1,850	1,590	1,480
5ft. 2in.	2,290	2,060	1,980	1,950	1,900	1,640	1,510
5ft. 3in.	2,350	2,100	2,030	2,000	1,950	1,690	1,550
5ft. 4in.	2,400	2,150	2,080	2,040	2,000	1,740	1,590
5ft. 5in.	2,460	2,200	2,140	2,080	2,050	1,780	1,640
5ft. 6in.	2,520	2,250	2,190	2,120	2,100	1,820	1,690
5ft. 7in.	2,570	2,300	2,240	2,160	2,150	1,860	1,730
5ft. 8in.	2,620	2,350	2,290	2,220	2,200	1,910	1,770
5ft. 9in.	2,680	2,400	2,340	2,260	2,250	1,950	1,800
5ft. 10in.	2,740	2,450	2,400	2,310	2,300	1,990	1,830
5ft. 11in.	2,800	2,500	2,450	2,360	2,350	2040	1,880
6ft. 0	2,860	2,550	2,500	2,410	2,400	2,090	1,930

*This intake will generally maintain normal weight at present level.

(These tables were developed by Norman Jolliffe, M.D., former Director of the Bureau of Nutrition, New York City Department of Health.)

Flexible Calorie Intake. The calorie needs in these tables are averages based on studies of women and men of normal weight engaged in moderate activities. Housewives, clerical workers and workers in light industry are included. NOTE: If you are in an occupation that requires heavy physical activity, you may be able to use anywhere from 10% to 20% more calories than indicated here, without gaining weight. If you are retired or homebound and not too active, your calorie needs could be 10% less than indicated.

Age Difference Considered. The tables also take into consideration the fact that younger people are more active generally. For example, 60% of the calorie requirements of those between 15 and 19 years of age are allocated for physical activity, and only 40% for persons past 60.

CALORIE COUNTING—AT A GLANCE

Here is a chart listing most common foods and their calorie counts. A little planning ahead will enable you to enjoy your favorite foods but with fewer calories.

---

1 cup equals 8 fluid ounces. 3 teaspoons (tsp.) equal 1 tablespoon (tbs.). 4 tablespoons (tbs.) equal ¼ cup.

Food and Measures	Approximate Calories
A	
Almonds . . 12-15	100
Apple butter . . 1 tbs.	40
Apples	
baked . . 1 lg. and 2 tbs. sugar	200
fresh . . 1 large	100
Applesauce, sweetened ½ cup.	100
Apricots	
canned in syrup . . 3 lg. halves and 2 tbs. juice	100
dried . . 10 sm. halves	100
Asparagus	
fresh or canned . . 5 stalks 5 ins. long	15
Avocado . . ½ pear 3½ x 3¼ ins.	185
B	
Bacon . . 2-3 long slices, cooked	100
Bacon fat . . 1 tbs.	100
Banana . . 1 med., 6 ins. long	90
Beans	
canned with pork ½ cup	175
dried . . ½ cup, cooked	135
lima, fresh or canned ½ cup	100
snap, fresh or canned ½ cup	25

Food and Measures	Approximate Calories
Beef (cooked)	
corned . . 1 slice 4 by 1½ by 1 ins.	100
dried . . 2 ozs.	100
hamburger . . 1 patty (3 ozs.)	300
round, lean . . 1 med. slice (2 ozs.)	125
sirloin, lean . . 1 av. slice (3 ozs.)	250
tongue . . 2 ozs.	125
Beet greens . . ½ cup, cooked	30
Beets, fresh or canned 2 beets 2 ins. in diam.	50
Biscuit, baking powder 2 ins. in diam.	100
Blackberries, fresh 1 cup	100
Blueberries, fresh 1 cup	90
Bologna . . 1 slice 2 ins. by ½ in. thick	100
Breads	
Boston brown . . 1 slice 3 ins. in diam. ¾ in. thick	90
corn (1-egg) 1-2 in. square	120
cracked wheat 1 slice, av.	80
dark rye . . 1 slice ½ in. thick	70
light rye . . 1 slice ½ in. thick	75

white, enriched
1 slice, av. 75

white, enriched
1 slice, thin 55

whole wheat, 60%
1 slice, av. 70

whole wheat, 100%
1 slice, av. 75

Broccoli . . 3 stalks
5½ ins. long 100

Brownies . . 1 piece 2 by
2 by ¾ ins. 140

Brussels sprouts
6 sprouts 1½ ins.
in diam. 50

Butter . . 1 tbs. 95

C

Cabbage, cooked . . ½ cup . . . 40
raw . . 1 cup 25

Cake
angel . . 1/10 of a lg.
cake 155

chocolate or vanilla,
no icing . . 1 piece
2 by 2 by 2 ins. 200

chocolate or vanilla, with
icing . . 1 piece
2 by 1½ by 2 ins. 200

cupcake with chocolate
icing . . 1 medium 250

Cantaloupe . . ½ of a 5½-in.
melon 50

Carrots . . 1 carrot 4 ins.
long 25

Cashew nuts . . 4-5 100

Cauliflower . . ¼ of a hd.
4½ ins. in diam. 25

Caviar . . 1 tbs. 25

Celery . . 2 stalks 15

Cheese
American cheddar
1 cube 1 1/8 ins. sq.
or 3 tbs. grated 110

cottage . . 5 tbs. 100
cream . . 2 tbs. 100

Cherries, sweet . . 15 lg. 75

Chicken
broiled . . ½ med.
broiler 270
roast . . 1 slice
4 by 2½ by ¼ ins. 100

Chinese cabbage
1 cup raw 20

Chocolate
milk, sweetened . . 1 oz. . . . 140
fudge . . 1 piece 1 in. sq.
by ¾ in. thick 100
malted milk . . fountain
size 460
mints . . 1 mint
1½ ins. in diam. 100
milk with almonds,
sweetened . . 1 oz. 150
syrup . . ¼ cup 200
unsweetened
1 square 160

Cider, sweet . . 1 cup 100

Clams . . 6 round 100

Cocoa, half milk, half
water . . 1 cup 150

Coconut . . ½ cup, fresh . . . 175

Cod-liver oil . . 1 tbs. 100

Cod steak . . 1 piece
3½ by 2 by 1 in. 100

Cola soft drinks
6-oz. bottle 75

Collards . . ½ cup, cooked . . . 50

Cooking fats, vegetable
1 tbs. 100

Corn . . ½ cup 70

Corn syrup . . 1 tbs. 75

Corn flakes . . 1 cup 80

Corn meal . . 1 tbs.,
uncooked 35

Cornstarch pudding
½ cup 200

Crackers

graham .. 1 square 35
peanut butter-cheese
    sandwich .. 1 cracker . . 45
round snack-type
    1 cracker 2 ins.
    in diam. 15
rye wafers .. 1 wafer 25
saltines .. 1 cracker
    2 ins. sq. 15
Cranberry sauce .. ¼ cup 100
Cream
    light .. 2 tbs. 65
    heavy .. 2 tbs. 120
    whipped .. 3 tbs. 100
Cream-puff shells .. 1 shell . . . 85
Cucumber .. ½ medium 10
Custard, boiled or baked
    ½ cup 130

D

Dates .. 4 100

E

Egg .. 1 medium size 75
Eggplant .. 3 slices 4 ins.
    in diam. ½ in. thick, raw . . . 50
Endive .. average serving 10
Escarole .. average
    serving 10

F

Figs, dried .. 3 small 100
Flour, white or whole grain
    1 tbs., unsifted 35
Frankfurter .. 1 sausage 125

G

Gelatin, fruit flavored, dry
    3 oz. pkg. 330
    ready to serve .. ½ cup . . . 85
Ginger ale .. 1 cup 85
Gingerbread, hot water
    2 by 2 by 2 ins. 200
Grapefruit .. ½ medium 50

Grapefruit juice,
    unsweetened .. 1 cup 100
Grape juice .. ½ cup 80
Grape nuts .. ¼ cup 100
Grapes
    American or Tokay
    1 bunch—22, av. 75
    seedless .. 1 bunch—
    30, av. 75
Griddle cakes
    1 cake 4 ins. in diam. 75

H

Halibut .. 1 piece
    3 by 1 3/8 by 1 ins. 100
Ham, lean .. 1 slice
    4¼ by 4 by ½ ins. 265
Hard sauce .. 1 tbs. 100
Hickory nuts .. 12-15 100
Hominy grits
    ¾ cup, cooked 100
Honey .. 1 tbs. 100

I

Ice cream .. ½ cup 200
Ice cream soda
    fountain size 325

J

Jellies and jams
    1 rounded tbs. 100

K

Kale .. ½ cup, cooked 50

L

Lamb, roast .. 1 slice 3½ by
    4½ by 1/8 ins. 100
Lard .. 1 tbs. 100
Lemon juice .. 1 tbs. 5
Lettuce .. 2 lg. leaves 5
Liver .. 1 slice
    3 by 3 by ½ ins. 100
Liverwurst .. 2 ozs. 130
Lobster meat .. 1 cup 150

M

Macaroni . . ¾ cup,
   cooked 100
Maple syrup . . 1 tbs. 70
Margarine . . 1 tbs. 100
Marshmallows . . 1 20
Milk
   buttermilk (fat-free)
    1 cup 85
   condensed . . 1½ tbs. 100
   evaporated . . ½ cup
    (1 cup diluted) 160
   instant non-fat dry
    6 tbs. 80
   skim milk, fresh
    1 cup 85
   whole milk . . 1 cup 170
   yogurt, plain
    1 cup 120-160
Mints, cream . .
   ½-in. cube 5
Molasses . . 1 tbs. 70
Muffins
   bran . . 1 medium 90
   1-egg . . 1 medium 130
Mushrooms . . 10 large 10
Mustard greens
   ½ cup, cooked 30

N

Noodles . . ¾ cup, cooked . . . 75

O

Oatmeal . . ¾ cup, cooked . . . 110
Oil—corn, cottonseed,
   olive, peanut, safflower
    1 tbs. 100
Okra . . 10-15 pods 50
Olives
   Green . . 4 medium, or
    3 extra large 15
   Ripe . . 3 small, or
    2 large 15
Onions . . 3-4 medium 100

Orange . . 1 medium 80
   juice . . 1 cup 125
Oysters . . 5 medium 100

P

Parsnips . . 1 parsnip
   7 ins. long 100
Peaches
   canned in syrup
    2 lg. halves and
    3 tbs. juice 100
   dried . . 4 medium
    halves 100
   fresh . . 1 medium 50
Peanut butter . . 1 tbs. 100
Peanuts, shelled . . 10 50
Pears
   canned in syrup
    3 halves and
    3 tbs. juice 100
   fresh . . 1 medium 50
Peas
   canned . . ½ cup 65
   fresh, shelled . . ¾ cup 100
Pecans . . 6 100
Pepper, green . . 1 medium . . . 20
Pickles, cucumber
   sour and dill . . 10 slices
    2 ins. in diam. 10
   sweet . . 1 small 10
Pies . . (sectors from
   9-in pies) apple
    3-in. sector 200
   lemon meringue
    3-in. sector 300
   mincemeat
    3-in. sector 300
   pumpkin . . 3-in. sector . . . 250
Pineapple
   canned, unsweetened
   1 slice ½-in. thick and
   1 tbs. juice 50
   fresh . . 1 slice
    ¾-in. thick. 50

juice, unsweetened

    1 cup 135

Plums

    canned . . 2 med. and

      1 tbs. juice 75

    fresh . . 2 medium 50

Popcorn, plain . . 1½ cups,

    popped 100

Popovers . . 1 popover 100

Pork chop, lean

    1 medium 200

Potato chips . . 8-10 large 100

Potato salad with

    mayonnaise . . ½ cup 200

Potatoes

    mashed . . ½ cup 100

    sweet . . ½ medium 100

    white . . 1 medium 100

Prune juice . . ½ cup 100

Prunes, dried . . 4 medium . . . 100

Pumpkin . . ½ cup 50

R

Radishes . . 5 10

Raisins . . ¼ cup 90

Raspberries, fresh . . 1 cup . . . 90

Rhubarb, stewed and

    sweetened . . ½ cup 100

Rice . . ¾ cup, cooked 100

Roll, Parker House

    1 medium 100

Rutabagas . . ½ cup 30

S

Salad dressing

    boiled . . 1 tbs. 25

    French . . 1 tbs. 90

    mayonnaise . . 1 tbs. 100

Salmon, canned . . ½ cup 100

Sardines, drained . . 5 fish

    3 ins. long 100

Sauerkraut . . ½ cup 15

Sherbet . . ½ cup 120

Soup, condensed . . 11-oz.

can

    mushroom 360

    noodle 290

    tomato 230

    vegetable 200

Spaghetti . . ¾ cup,

    cooked 100

Spinach . . ½ cup, cooked . . . 20

Squash

    summer . . ½ cup, cooked . . 20

    winter . . ½ cup, cooked . . . 50

Strawberries, fresh . . 1 cup . . . 90

Sugar

    brown . . 1 tbs. 50

    granulated . . 1 tbs. 50

    powdered . . 1 tbs. 40

Sweetbreads, calves'

    1 pair med.-sized 200

Swiss chard

    ½ cup leaves and stems . . . 30

T

Tangerines . . 1 medium 60

Tapioca, uncooked . . 1 tbs. . . 50

Tomato juice . . 1 cup 60

Tomatoes, canned . . ½ cup . . . 25

    fresh . . 1 medium 30

Tuna fish, canned

    ¼ cup, drained 100

Turkey, lean . . 1 slice

    4 by 2½ by ¼ ins. 100

Turnip . . 1 turnip

    1¾ ins. in diam. 25

Turnip greens . . ½ cup,

    cooked 30

V

Veal, roast . . 1 slice

    3 by 3¾ by ½ ins. 120

W

Waffles . . 1 waffle

    6 ins. in diam. 250

Walnuts . . 8 100

Watermelon .. 1 round slice
   6 ins. in diam.
      1½ ins. thick, no rind . . . 190
Wheat
   flakes .. ¾ cup 100
   germ .. 1 tbs. 25
   shredded .. 1 biscuit 100

* * *

Alcoholic beverages
   beer .. 8 ozs. 120

gin .. 1½ ozs. 120
rum .. 1½ ozs. 150
whiskey .. 1½ ozs. 150

Wines
   champagne .. 4 ozs. 120
   port .. 1 oz. 50
   sherry .. 1 oz. 40
   table, red or white
      4 ozs. 95

YOUR HIGH-ENZYME AND LOW-CALORIE DAILY DIET PLAN

The foods in the following suggested daily diets may be eaten any time during the day, in varied combinations. For example, you may find it more satisfying to divide this 3-meal plan into 4 or 5 smaller meals. And perhaps you prefer to take part—or all—of your milk at breakfast on your cereal or in your beverage.

If you wish, have your vegetables in a tossed salad with a fruit juice dressing for enzyme boosting. The only rule to remember is this: Count everything you eat and keep your total daily food intake within the bounds of this diet plan.

Raw Foods Are A Must. All fruits and palatable vegetables must be eaten raw. This will give your body a daily supply of enzymes that will help metabolize body fat and calories and help keep you permanently slim. Each of these meals features a raw fruit or vegetable or both. This provides you with a high-enzyme daily intake, together with a low-calorie plan. Most important, you enjoy most of your favorite foods and can lose weight while you eat.

1000 CALORIE ENZYME DAILY DIET*

Breakfast

Fresh fruit or juice . . . . 1 serving—½ cup
Egg—cooked without fat 1
    or
Cereal . . . . 1 small serving
Bread . . . . 1 slice
Butter or margarine . . . . 1 level teaspoon
Skim milk or buttermilk . . . . 1 glass—6 ounces

Dinner

Lean meat, fish, or poultry . . . . 3 ounces (cooked)
Vegetables (raw) . . . . ½ cup

Skim milk or buttermilk . . . . 1 glass—6 ounces
Fruit (raw) . . . . 1 serving—½ cup

Lunch or Supper

Cottage cheese or lean meat . . . . ½ cup of cheese or
                                                       2 ounces of meat
Vegetables (raw) . . . . ½ cup
Skim milk or buttermilk . . . . 1 glass—6 ounces
Fruit (raw) . . . . 1 serving—½ cup

*Approximate

1200 CALORIE ENZYME DAILY DIET*

Breakfast

Fresh fruit or juice . . . . 1 serving—½ cup
Egg—cooked without fat 1
    or
Cereal . . . . 1 small serving
Bread . . . . 1 slice
Butter or margarine . . . . 1 level teaspoon
Skim milk . . . . 1 glass—6 ounces

Dinner

Lean meat, fish, or poultry . . . . 4 ounces (cooked)
Vegetables (raw) . . . . ½ cup
Potato or bread . . . . 1 small potato or
                                                       1 slice of bread
Butter or margarine . . . . 1 level teaspoon
Skim milk . . . . 1 glass—6 ounces
Fruit (raw) . . . . 1 serving—½ cup

Lunch or Supper

Cottage cheese or lean meat . . . . ½ cup of cheese or
                                                       2 ounces of meat
Vegetables (raw) . . . . ½ cup
Skim milk . . . . 1 glass—6 ounces
Fruit (raw) . . . . 1 serving—½ cup

*Approximate

1500 CALORIE ENZYME DAILY DIET

Breakfast

Fresh fruit or juice . . . . 1 serving—½ cup

Egg—cooked without fat 1
    or
Cereal . . . . 1 serving of cereal
                                   (1 cup, prepared,
                                   or ½ cup, cooked)
Bread . . . . 1 slice
Butter or margarine . . . . 1 level teaspoon
Skim milk . . . . 1 cup—8 ounces

Dinner

Lean meat, fish, or poultry . . . . 4 ounces (cooked)
Vegetables (raw) . . . . ½ cup
Potato . . . . 1 small
Butter or margarine . . . . 1 level teaspoon
Skim milk . . . . 1 cup—8 ounces
Fruit (raw) . . . . 1 serving—½ cup

Lunch or Supper

Cottage cheese or lean meat . . . . ½ cup of cheese or
                                   2 ounces of meat
Vegetables (raw) . . . . ½ cup
Bread . . . . 1 slice
Butter or margarine . . . . 1 level teaspoon
Skim milk . . . . 1 cup—8 ounces
Fruit, plain custard, or
    plain cookies . . . . ½ cup of fruit or
                                   custard, or 2 cookies

*Approximate

If you prefer a:

PACKED LUNCH For 1200 Calorie Enzyme Daily Diet*

Breakfast	**Packed Lunch**	**Dinner**
½ cup fruit	Sandwich:	4 oz. meat, fish or
1 egg or cereal	2 thin slices bread	poultry
1 slice toast	1 teaspoon mayonnaise	½ cup cooked
1 teaspoon butter or	2 oz. lean meat,	vegetable
margarine	fish or poultry	raw vegetables, freely
1 glass skim milk	Wedge of raw cabbage	1 glass skim milk
	Whole raw carrot	½ cup fruit—fresh
	Fresh fruit	
	1 glass skim milk	

½ cup fruit	Sandwich:	4 oz. meat, fish or
1 egg or cereal	2 thin slices bread	poultry
1 slice toast	2 oz. lean meat,	½ cup raw
1 teaspoon butter or	fish or poultry	vegetable
margarine	1 teaspoon mayonnaise	raw vegetables, freely
1 glass skim milk	lettuce	1 glass skim milk
	2 stalks celery	½ cup fruit—fresh
	1 small cucumber	
	Fresh fruit	
	1 glass skim milk	

½ cup fruit	Sandwich:	4 oz. meat, fish or
1 egg or cereal	2 thin slices bread	poultry
1 slice toast	2 oz. chicken or	½ cup cooked
1 teaspoon butter	meat chopped and	vegetable
1 glass skim milk	mixed with 1 table-	raw vegetables,
	spoon chopped celery	freely
	and a little chopped	1 glass skim milk
	onion	½ cup fruit—fresh
	Large fresh tomato	
	½ green pepper	
	Fruit	
	1 glass skim milk	

* Approximate

HOW TO ENJOY SWEETS THAT OFFER AN ENZYME-CATALYST ACTION ON CALORIES

Sugar is a pure carbohydrate substance that offers no nutritive or health building values. Sugar offers only one substance—*calories*. Plenty of them, too! But many folks are conditioned to the enjoyment of sweets. An excessive amount of sugar offers high amounts of calories and weight builds up. To help ease and eliminate the craving for sugar, begin by enjoying more healthful sweets that also contain enzymes. The enzymes can then create a catalyst action on the calories so they can be metabolized swiftly.

Safe, Healthful Sweets. Natural sweets include raw honey, honey comb, blackstrap or Barbados molasses, Big "GG" syrup made from unrefined cane syrup, and 100% Pure Maple Syrup or Sugar. Try natural date sugar, carob powder, dried banana flakes as *substitutes* for refined sugar.

How Natural Sweets Help Metabolize Calories. These natural sweets are unrefined and non-processed and are storehouses of enzymes that create a ˙catalyst action on calories so they are metabolized and "burned" instead of being stored in the adipose (fat) cells. Enzymes in these natural sweets are used by the metabolic system to trigger additional body-heat production (burning calories) to create the ECD action that cleanses and "washes" out calories from the cell tissues. Therefore, with this Enzyme-Catalyst Diet principle in activity, you can enjoy *healthful* sweets that will metabolize calories. It is like having your sweets and eating them, too!

How "Sugarholic" Indulges In Sweets That Keep Him Permanently Slim. Michael N. was a "sugarholic" ever since early childhood. His parents rewarded him for being good with candy, cookies, sweets, cakes, pies, pastries, ice cream, confections, etc. By the time he was a teen-ager, Michael N. was more than 75 pounds overweight. He was burdened with unsightly fat. He had heavy "breasts," a barrel-like stomach, ballooning buttocks and thick, flabby thighs. When he reached the age of 22, he had a triple chin, heavy jowls and layers upon layers of corpulent flab. He was the object of ridicule in his neighborhood. If he had to walk a few blocks, he was always out of breath. He wheezed, coughed, sputtered. His parents no longer gave him sweets for being good, but the habit had taken hold of him. He was a "sugarholic" and had an obsessive craving for sweets. When he was told that he was risking his life, he became desperate. He was told to continue indulging in sweets . . . but *natural sweets.* Here is what he did:

1. He drank beverages such as skim milk, herb tea, flavored with raw honey.
2. Instead of jam on bread, he would use honeycomb on a thinner slice of whole grain bread.
3. Over whole grain cereal, he would use pure date sugar.
4. Whenever he wanted to have a sweet, he reached for a slice of fresh fruit, or banana and apple slices sweetened with a bit of molasses.

Benefits: These were natural sweets that contained enzymes which acted as catalysts to metabolize calories in these foods so that weight could be controlled.

Michael N. could indulge in his sweet urge, but now he stored up fewer and fewer calories. He also found that he was satisfied with *less quantities* and *reduced portions* of sweets. A gradual reduction slowly brought down his weight. Natural sweets have a nutritive benefit with their vitamins and minerals and these entered into the catalyst action to create better oxidation in the cells and gradual calorie countdown. Soon, Michael N. could shed some 60 pounds. He was soon slim and youthful looking. More important, he had conquered his sugar-eating urge and felt cured forever. The simple Enzyme-Catalyst Diet program that changed refined sugars to natural sugars had made him lose his excess weight permanently, while he enjoyed sweets!

ANTI-ENZYME ACTION OF REFINED SUGAR

To help keep healthy and slim, you should avoid refined sugar. It interferes with the metabolic system and hinders the enzymatic catalyst action of body rhythm. Sugar also causes an abnormal production of many hormones, especially hormones. This hormonal imbalance can cause diabetes, most likely by overworking the insulin-producing cells of the pancreas glands.

Causes Rise In Blood Fats. Refined sugar, causing overproduction of insulin, raises blood levels of fats (called triglycerides) which are deposited in blood vessel walls. This causes weight buildup that can also contribute to heart attacks and strokes. Furthermore, excessive refined sugar impairs natural body enzyme action so that metabolism is slowed down. This may create "hypertriglyceridemia", or high blood levels of fats. This is unhealthy insofar as weight and heart health is concerned. By inhibiting and destroying enzymes, sugar is a serious threat to body life.

"HIDDEN SUGARS" IN EVERYDAY FOODS [2]

You may say you don't eat much sugar. But read the following chart. Note that many everyday foods, especially refined and processed foods, have "hidden sugars." In the size portion column, calculate 2 tablespoons for 1 fluid ounce. *Example:* 2 tablespoons of fudge give you 4 1/2 teaspoons of sugar. *Calorie Count:* 3 teaspoons of white sugar will give you 40 calories and 11 carbohydrates. With

---

[2]Courtesy *Parker Natural Health Bulletin,* West Nyack, New York 10994. Vol. 4, No. 21, October 14, 1974. Available by subscription.

these figures at hand, you can see how many calories you eat with these everyday foods that contain "hidden sugars."

FOOD ITEM	SIZE PORTION	APPROXIMATE SUGAR CONTENT IN TEASPOONFULS OF GRANULATED SUGAR
BEVERAGES		
COLA DRINKS	1 (6 oz. bottle or glass)	3½
CORDIALS	1 (¾ oz. glass)	1½
GINGER ALE	6 oz.	5
HI-BALL	1 (6 oz. glass)	2½
ORANGE-ADE	1 (8 oz. glass)	5
ROOT BEER	1 (10 oz. bottle)	4½
SEVEN-UP	1 (6 oz. bottle or glass)	3¾
SODA POP	1 (8 oz. bottle)	5
SWEET CIDER	1 cup	6
WHISKEY SOUR	1 (3 oz. glass)	1½
CAKES & COOKIES		
ANGEL FOOD	1 (4 oz. piece)	7
APPLE SAUCE CAKE	1 (4 oz. piece)	5½
BANANA CAKE	1 (2 oz. piece)	2
CHEESE CAKE	1 (4 oz. piece)	2
CHOCOLATE CAKE (Plain)	1 (4 oz. piece)	6
CHOCOLATE CAKE (Iced)	1 (4 oz. piece)	10
COFFEE CAKE	1 (4 oz. piece)	4½
CUP CAKE (Iced)	1	6
FRUIT CAKE	1 (4 oz. piece)	5
JELLY-ROLL	1 (2 oz. piece)	2½
ORANGE CAKE	1 (4 oz. piece)	4

FOOD ITEM	SIZE PORTION	APPROXIMATE SUGAR CONTENT IN TEASPOONFULS OF GRANULATED SUGAR
POUND CAKE	1 (4 oz. piece)	5
SPONGE CAKE	1 (1 oz. piece)	2
STRAWBERRY SHORTCAKE	1 serving	4
BROWNIES (Unfrosted)	1 (¾ oz.)	3
CHOCOLATE COOKIES	1	1½
FIG NEWTONS	1	5
GINGER SNAPS	1	3
MACAROONS	1	6
NUT COOKIES	1	1½
OATMEAL COOKIES	1	2
SUGAR COOKIES	1	1½
CHOCOLATE ECLAIR	1	7
CREAM PUFF	1	2
DONUT (plain)	1	3
DONUT (Glazed)	1	6
SNAIL	1 (4 oz. piece)	4½

CANDIES

FOOD ITEM	SIZE PORTION	APPROXIMATE SUGAR CONTENT IN TEASPOONFULS OF GRANULATED SUGAR
AVERAGE CHOCOLATE MILK BAR (example: Hershey bar)	1 (1½ oz.)	2½
CHEWING GUM	1 stick	½
CHOCOLATE CREAM	1 piece	2
BUTTERSCOTCH CHEW	1 piece	1
CHOCOLATE MINTS	1 piece	2
FUDGE	1 oz. square	4½
GUM DROP	1	2
HARD CANDY	4 oz.	20
LIFESAVERS	1	1/3
PEANUT BRITTLE	1 oz.	3½

FOOD ITEM	SIZE PORTION	APPROXIMATE SUGAR CONTENT IN TEASPOONFULS OF GRANULATED SUGAR
CANNED FRUITS & JUICES		
CANNED APRICOTS	4 halves and 1 Tbsp. syrup	3½
CANNED FRUIT JUICES (Sweetened)	½ cup	2
CANNED PEACHES	2 halves and 1 Tbsp. syrup	3½
FRUIT SALAD	½ cup	3½
FRUIT SYRUP	2 Tbsp.	2½
STEWED FRUITS	½ cup	2
DAIRY PRODUCTS		
ICE CREAM	1/3 pt. (3½ oz.)	3½
ICE CREAM BAR	1	1-7 depending on size
ICE CREAM CONE	1	3½
ICE CREAM SODA	1	5
ICE CREAM SUNDAE	1	7
MALTED MILK SHAKE	1 (10 oz. glass)	5
JAMS & JELLIES		
APPLE BUTTER	1 Tbsp.	1
JELLY	1 Tbsp.	4-6
ORANGE MARMALADE	1 Tbsp.	4-6
PEACH BUTTER	1 Tbsp.	1
STRAWBERRY JAM	1 Tbsp.	4
DESSERTS, MISCELLANEOUS		
APPLE COBBLER	½ cup	3
BLUEBERRY COBBLER	½ cup	3
CUSTARD	½ cup	2

FOOD ITEM	SIZE PORTION	APPROXIMATE SUGAR CONTENT IN TEASPOONFULS OF GRANULATED SUGAR
FRENCH PASTRY	1 (4 oz. piece)	5
JELLO	½ cup	4½
APPLE PIE	1 slice (average	7
APRICOT PIE	1 slice	7
BERRY PIE	1 slice	10
BUTTERSCOTCH PIE	1 slice	4
CHERRY PIE	1 slice	10
CREAM PIE	1 slice	4
LEMON PIE	1 slice	7
MINCEMEAT PIE	1 slice	4
PEACH PIE	1 slice	7
PRUNE PIE	1 slice	6
PUMPKIN PIE	1 slice	5
RHUBARB PIE	1 slice	4
BANANA PUDDING	½ cup	2
BREAD PUDDING	½ cup	1½
CHOCOLATE PUDDING	½ cup	4
CORNSTARCH PUDDING	½ cup	2½
DATE PUDDING	½ cup	7
FIG PUDDING	½ cup	7
GRAPENUT PUDDING	½ cup	2
PLUM PUDDING	½ cup	4
RICE PUDDING	½ cup	5
TAPIOCA PUDDING	½ cup	3
BERRY TART	1	10
BLANC-MANGE	½ cup	5
BROWN BETTY	½ cup	3
PLAIN PASTRY	1 (4 oz. piece)	3
SHERBET	½ cup	9

FOOD ITEM	SIZE PORTION	APPROXIMATE SUGAR CONTENT IN TEASPOONFULS OF GRANULATED SUGAR
SYRUPS, SUGARS & ICINGS		
BROWN SUGAR	1 Tbsp.	3 (actual sugar content)
CHOCOLATE ICING	1 oz.	5
CHOCOLATE SAUCE	1 Tbsp.	3½
CORN SYRUP	1 Tbsp.	3 (actual sugar content)
GRANULATED SUGAR	1 Tbsp.	3 (actual sugar content)
HONEY	1 Tbsp.	3 (actual sugar content)
KARO SYRUP	1 Tbsp.	3 (actual sugar content)
MAPLE SUGAR	1 Tbsp.	5 (actual sugar content)
MOLASSES	1 Tbsp.	3½ (actual sugar content)
WHITE ICING	1 oz.	5

Found In Processed Foods. Sugar in one form or another, may be called dextrose, glucose, corn syrup, corn sugar or invert sugar. It is still anti-enzyme refined sugar. It is found in canned soups (bouillon cubes, too), in many cheese spreads, dinner rolls, luncheon meats such as bologna and pastrami, canned Welsh rarebit, artificially flavored and colored peanut spread, mustard, canned foods. Also found in seasoned rice mixes, most add-the-main-ingredient packages, stuffed mixes for clam dip, hamburger, garlic spread, poultry coating, many frozen main dishes, including pizza. Sugar is also found in baby foods. One good rule: *read the label* of processed and packaged foods. If it contains sugar by its name or other names as listed above, pass it up. If no ingredients are given, it may contain sugar, added to create flavor that processing destroys, so pass it up, too.

HOW A FAT FAMILY LOST WEIGHT ON A
TASTY-SWEET BUT SUGAR-LOW DIET

Fat did not necessarily run in the family of Rose Y. Rather, much of what they ate was so sugar-flavored, the entire family was fat. To lose weight, Rose Y. followed this tasty-sweet Enzyme-Catalyst Diet Program that was sugar-low and enzyme high:

1. In cooking, Rose Y. replaced much of sugar with pure maple syrup. One cup of sugar equals two cups of maple syrup. Do not replace more than half the sugar. Reduce the liquid in the recipe by one-quarter cup. **ECD Benefit:** The maple syrup contained vitamins and minerals used by enzymes as a buffering action against calorie buildup in the adipose cell tissues.

2. Honey, with its distinctive flavor, was substituted for sugar in cookies and cakes. Just use 7/8th cup of honey for each 1 cup of sugar. (In breads, one cup of honey equals one cup of sugar.) Reduce liquid by three tablespoons. **ECD Benefit:** Enzymes take the minerals from the honey to create better metabolism and combustion of calories to control weight.

3. Use molasses and maple syrup in recipes specifically designed for their use. **ECD Benefit:** These natural sweeteners do not block enzymatic action and calories can be metabolized with little interference as is the problem with refined sugar.

4. Reduce sugar in most baking recipes. The product will have more texture, more compact or pound-cake form. **ECD Benefit:** Reduced sugar still offers sweet taste but less sugar blockage in the system, so enzymes can function freely. As you progress, keep reducing sugar until entirely eliminated.

5. Replace sugar in most cooked dishes with dried fruits. Try them in relishes and sweet-and-sour dishes. **ECD Benefit:** Sun-dried fruits are powerhouses of enzymes which can help metabolize body calories to control weight while still giving you a sweet taste.

6. Use fresh fruit slices or raisins on top of cereals for natural sweetness. **ECD Benefit:** Fresh fruit contains vitamins and minerals which energize enzymes and whip them up into speedy metabolism of calories.

7. Sweeten plain yogurt with fresh fruit. **ECD Benefit:** A fermented food, yogurt is brimming with enzymes that take up vitamins from the fruit and promote cellular metabolism and weight control.

8. Enjoy a gelatin dessert by combining unflavored gelatin with fresh fruit juices. Sprinkle one package of gelatin on one-quarter cup' of fruit juice to soften. Now bring 1 3/4 cups fruit juice to a boil and stir in softened gelatin. Chill until set. Try with high-enzyme orange, grape, apple or pineapple juice. **ECD Benefit:** Enzymes from juices combine with the protein in the gelatin and create a slow and steady "flame" to help burn up calories.

9. If you must make cookie-crust recipes, do NOT add any sugar. Those recipes that use chocolate and vanilla wafers, graham crackers or ginger snaps are already sugary. If possible, substitute with fresh enzyme-high fruit slices. **ECD Benefit:** If you must eat these cookies, do so with fresh fruit that helps guard against calorie buildup. Fruit enzymes will metabolize much of the calories.

10. For dessert, satisfy your sweet tooth with fresh fruit slices. Also try cheese with fruit. **ECD Benefit:** Enzymes in the fruit become activated and amplified by the protein in the cheese to promote better calorie metabolism. **Tip:** Use a strong flavored cheese with a tart fruit. Use a mild or delicate cheese with a sweet fruit.

Example: Rose Y. trained her "fat family" to enjoy tasty-sweet and enzyme-high desserts such as: apples with blue cheeses or Cheddar; pears with provolone; pineapple slices with camembert or brie; bananas with Swiss cheese.

Pounds Melt Before Their Eyes. Rose Y. and her family started to slim down on a high-enzyme and sugar-free program so that they could actually see pounds melting before their eyes. Every day, they looked at one another, remarking how slim each was getting to be. Within a short time, the "fat family" became permanently slim. Enzymes had made them a new and healthier family . . . while they ate sweets!

Sugar from refined sources are villains in the weight control goal. Sugar has a blocking action upon enzymes and slows down their weight-melting functions. Control and eliminated refined sugar and your body metabolism can function efficiently and smoothly to help protect against excessive weight build-up.

Special Points:

1. Enzymes create a high osmotic pressure in which accumulated calories in the cell tissues are burned up and melted down.

2. Begin by planning your calorie-burning program, based upon the official U.S. government recommendations. Simple 5-step plan.

3. Keep a calorie diary to chart your weight gain and loss.

4. Plan the amount of daily calories you need for good health. Overweight begins by setting a target for this daily amount. Anything over that amount causes weight buildup.

5. Count calories at-a-glance with the handy chart.

6. Raw foods offer enzymes that can combat calorie buildup. Enjoy any of the calorie enzyme diets presented. Eat and slim down.

7. Barbara U. shed over 40 excess pounds on an easy Enzyme-Catalyst Diet Program.

8. Michael N. was a "sugarholic" but used enzymes to slim down youthfully.

9. Note "hidden sugars" and keep these items to a minimum in your daily diet.

10. Rose Y. slimmed down her "fat family," including herself on a 10-step tasty-sweet and enzyme-high program.

7

The "Stay Slim Forever" Power of Fruit Enzymes

Steve McK. had a big appetite. He loved the juicy good taste of broiled steaks, thrilled to the luscious goodness of lamb chops, looked forward to hearty casseroles and frequent goulash meals. When he started to follow the Enzyme-Catalyst Diet Program, he was some 50 pounds overweight. As he began eating more and more raw fruits, before, during and after his meals, he started to shed more pounds. Enzymes in the raw fruits improved the catalyst action upon calorie-building fat and carbohydrate foods. But Steve McK. worked long hours and would often come home long after the food markets were closed. His wife worked late, too. This meant they did not always have fresh fruit available. They had to stock up in advance. So they began buying canned fruits.

Weight Mounts Up On Canned Fruits. Steve McK. started to put on more and more weight as he used canned fruits instead of fresh fruits. Soon, he was so heavy, he had to buy new clothes. His waistline was an unsightly 44, and still growing! Something was wrong. He ate fruit but was still gaining more weight. His wife switched to water-packed fruit, instead of sugar syrup packed. This reduced caloric intake, but Steve McK. still kept gaining weight. Then his wife changed her work schedule so she would have time to stop off at a produce market for fresh, raw fruit. As soon as he began eating more and more fresh, raw fruit, Steve McK. started to lose his heavy weight. His waistline trimmed down to 34. He could use his discarded clothes again. His scales showed a loss of some 44 pounds. He was now down to 175 and was slowly losing more excess weight. The secret? *Fresh, raw fruits are prime sources of cell-shrinking enzymes.* Canned fruits, whether water or sugar syrup packed, are devoid of enzymes. The canning process calls for pre-cooking and heating and this destroys valuable enzymes. To "stay slim forever," fresh, raw fruit is the key to a permanent weight loss. Steve McK. learned the hard way, but the lesson was well taught. Now he makes it a rule that fruit must be fresh and raw . . . or he won't eat it at all since canned fruit is "dead" fruit with "dead" enzymes.

HOW RAW FRUIT ENZYMES SLIM DOWN BODY CELLS

When you eat a selection of fresh, raw fruit, its enzymes alert your metabolism to stimulate your body to an increased energy that will "scrub" the cells and "burn up" excessive fat accumulation. Raw fruit enzymes promote a self-digestion of eaten food within the intestinal tract, thereby relieving the work of the digestive glands. These digestive glands can now work full force in using enzymes for the oxidation of accumulated fats, carbohydrates and calories that have expanded the cells and caused weight buildup.

Chain Reaction Causes "Stay Slim Forever" Benefit. Raw fruit enzymes work without interference and, by oxygen fixation, establish a catalyst action in the intestinal tract to cause multiplication of a beneficial coli bacteria. This substance drives away weight-building calories. It creates an internal balance that protects the body against excessive cellular obesity, the basic cause of overweight.

Creates Protective Barrier Against Fat Buildup. Raw fruit enzymes use a catalyst action to create a protective barrier against *digestive leucocytosis,* or fat buildup. Leucocytosis is a mobilization of certain blood substances that tend to raise fat levels and enlarge or "fatten" the nucleus of the cells. But eating raw fruits will send a supply of enzymes into the digestive system to protect against *digestive leucocytosis.* This keeps the cells from becoming too weighty. The body is kept slim, too.

How A Secretary Melted Down "Spreading Hips." Stella D., as a secretary, would sit while taking dictation or doing her typing. This caused her "spreading hips" or so-called "secretary's bottom." It was unflattering to her figure. Also, she disliked rigorous dieting. She did not want to deny herself her favorite foods. Since she frequently traveled, she could not stick to a diet for a long time. She wanted to take off those excess pounds while enjoying most of her favorite foods. But she wanted to keep those pounds off . . . for good. She ate fruits, but only when the mood hit her. She was told about the Enzyme-Catalyst Diet Program and instructed on using this very simple method:

Begin each meal with raw fruit. If the meal consists of cooked food, begin it with raw fruit and then end it with raw fruit.

Stella D. followed this simple ECD plan, and soon found that not only did her appetite become self-controlled, but her "secretary's bottom" was slimming down. Soon, she had a neat, curvaceous figure again. Now she uses this easy ECD Program daily, even when traveling, and has little weight problems, if any.

Secret Of "Stay Slim Forever" Power Of Raw Fruit Enzymes. Enzymes in raw fruit at the *beginning* of a meal have free rein in the digestive system and can work without interference in creating what is called *a micro-electric reaction in the core of the cell tissues.* Once this micro-electric reaction is created, cell respiration is heightened. Cell metabolism is stimulated. The powers of resistance and cell renewal increase. Body metabolism is alerted. This creates a chain reaction. All efforts are beamed at the accumulated fats, carbohydrates and calories that are lodged within the membranes and core of the cells. *The micro-electric reaction therefore will "break up" and dissolve these fat-causing substances, bringing about a time-release action to cause around-the-clock weight control.* The secret is to use raw fruits *before* your meal. You may finish your meal with more raw fruits for double-action enzyme power to perpetuate this micro-electric reaction that works within the core of the cell where the cause of fat is corrected. You can help "stay slim forever" with this easy and tasty method.

RAW FRUIT ENZYMES WORK QUICKLY TO PROMOTE WEIGHT LOSS

What Is Fruit? Botanically, fruit is the edible part of a plant that results from the development of pollinated flowers, such as oranges, apples, peaches, plums, etc. Fruit is the edible capsule that surrounds the seed.

Nature-Created Enzymes. During the growing process of fruits, Nature uses sunshine and oxygen to convert starch into fruit sugars and to boost the vigor of the enzymes. While under the ripening influence of the sun, the enzymes are strengthened. They are "pre-cooked," so to speak. When you eat a luscious raw fruit, you eat the Nature-prepared enzymes which are energized by *levulose* (fruit sugar) to travel throughout the body, to penetrate the cells, to create oxidation or a burning of the calories so that weight loss can take place. All this is done quickly because Nature has done this pre-digestion of starch, converting it into *levulose,* an enzyme-activator. This *levulose* is a substance that is completely digested. It is ready for speedy absorption and assimilation. It works with lightning swift action upon enzymes which are speedily sent to the billions of cells where they can perform their weight-losing action.

Finally, the natural carbons contained in the fruit, spark the enzymes to generate the cerebro-vital reactions that will alert the metabolism to burn up excessive calories.

It is the natural way to promote steady weight control. It works swiftly, even while you sleep. It takes weight off . . . and it keeps weight off. *Eating raw fruit before and after a meal can give you this miracle weight loss reaction.*

THESE FRUIT ENZYMES HELP MELT FAT-CAUSED POUNDS

Before eating a heavy meat meal, enjoy a plate of strong enzyme fruits such as: oranges, grapefruit wedges, tangerines with a sprinkle of lemon and lime juice, gooseberries, currants, pineapple slices. **ECD Benefit:** Enzymes in these fruits take up the tartaric acid and work to help digest and pierce the strong fat components of eaten meats, breaking down the connective tissues so that the body cells are spared fat buildup. They help melt fat-caused pounds.

THESE FRUIT ENZYMES HELP MELT
CARBOHYDRATE-CAUSED POUNDS

Before eating a heavy carbohydrate or starchy meal such as spaghetti, noodles, casseroles, bread products, enjoy a plate of strong enzyme fruits as: apples, pears, currants, berries, grapes, cherries, persimmons, plums, cherries. **ECD Benefit:** Enzymes in these fruits take up the malic acid and use these to transform carbohydrates into better metabolized sugar which can then be better catalyzed and protected against cellular buildup.

THESE FRUIT ENZYMES HELP MELT
CALORIE-CAUSED POUNDS

Before eating a heavy calorie meal such as excessive sweets, enjoy a plate of strong enzyme fruits as: raspberries, grapes, currants, plums, oranges, bananas, dates, figs, raisins, papayas. **ECD Benefit:** Enzymes in these fruits take up their oxalic acid and use these to metabolize calories, helping them to become oxidated so they will not cling to the cell nuclei where they cause weight increase.

Enjoy ECD Slimming Program. Because fruits are so luscious, they let you enjoy the Enzyme-Catalyst Diet Program to keep you slim forever.

U.S. GOVERNMENT SUGGESTIONS ON USING ENZYME FRUITS

Here are the official U.S. Government recommendations for using enzyme fruits[1] in everyday eating and helping to keep slim:

How To Prepare Fresh Fruits

1. Wash fresh fruits thoroughly, whether you serve them raw or cooked. Wash berries in a colander under gently running cold water. Trim away small bruises and injured areas; discard fruits that are too soft or decayed.

2. When you pare apples and pears, make parings as thin as possible. The skin of ripe peaches and apricots sometimes adheres tightly. To loosen it for peeling, dip the fruit into boiling water for about 45 seconds, then into cold water. You can then grasp the loosened skin of the fruit. Peel gently, using the dull edge of a knife.

3. To prepare orange or grapefruit halves, run a sharp knife around each section to loosen it from the membrane and skin. If you peel the whole fruit, remove all of the inner peel with a sharp knife.

4. Some fruits—apples, peaches, pears and bananas—turn brown if allowed to stand after they have been cut or peeled. To minimize discoloration, dip these fruits into citrus fruit juice (lemon, lime, orange or grapefruit) or pineapple juice. (This also increases their enzyme effectiveness.)

SIX WAYS TO SERVE ENZYME FRUITS

Here are the six official U.S. Government recommendations for serving enzyme fruits:

1. *Fruit Appetizers.* Combine several fresh fruits in a fruit cup. Or try a combination of fresh and frozen fruits.

---

[1] *Fruits In Family Meals,* Home and Garden Bulletin No. 125, U.S. Department of Agriculture, 1968, Washington, D.C.

2. *Fruit Salads.* Arrange fresh fruits, or a mixture of fresh or frozen fruits, on crisp greens. Serve with a tangy fruit salad dressing, or a mixture of apple cider vinegar and oil (for greater enzyme vigor). Try combining fruits with other foods—crisp raw vegetables, cooked meats and poultry, cheese, nuts and cooked seafood.

3. *Fruit Plates.* Arrange several fruits on a bed of crisp greens; add cheese or sherbet. Serve with small sandwiches or a hot bread.

4. *Fruit Garnishes.* A fresh fruit garnish makes many a main dish, salad or dessert more appetizing. It is an excellent way to provide living enzymes for cell metabolism. Fruit enzyme garnishes that go well with meats and poultry are: Whole cranberry sauce in orange cups, peach halves, apple rings, pineapple spears and orange slices. With appetizers, salads and desserts, try thin slices of lemon and lime, melon balls, large whole strawberries and green grapes.

5. *Fruit Snacks.* Choose fresh fruits to eat alone or with snacks, milk, cheese. For an Enzyme-Catalyst Diet trick—serve fresh fruit such as an apple or an orange, with a concentrated sweet such as a confection if that is your compulsion. The enzymes will help metabolize the calories from the sweet and guard against weight buildup.

6. *Fruit Desserts.* Serve fresh raw fruits singly or in combination, sweetened with a bit of honey, plain or topped with milk, low-fat cream, yogurt, skim milk cottage cheese or any cheese slices made from skim milk.

FRUIT ENZYME APPETIZERS IN MINUTES

Here are three official U.S. Government recommendations for serving and enjoying enzyme appetizers:

1. Dip banana chunks in lemon juice and roll in finely chopped nuts. Spear on toothpicks.

2. Dip unpared apple rings and pear wedges in lemon juice and spread with a mixture of Roquefort or blue cheese and cottage cheese made from skim milk.

3. String on toothpicks two or more of the following: Fresh pineapple cubes, seedless grapes, whole fresh berries, pear and apple chunks (dipped in lemon juice), cantaloup cubes,

orange sections. Serve with yogurt or blue cheese dip or try cottage cheese made from skim milk as a tasty dip.

ECD Benefits: These U.S. Government recommendations for using enzyme fruits will give your body the needed enzymes for better cellular metabolism so that there is improved micro-electric reaction in the nucleus to promote weight control and weight loss. Eat fruits daily and the enzyme-catalyst method will keep weight off *permanently!*

HOW THREE EVERYDAY FRUITS GAVE HOUSEWIFE A MOVIE STAR FIGURE

Evelyn E. was always sampling and tasting foods that cooked on the stove. It was this snacking that put on so much weight. Her dress size increased. She found it difficult to get up from chairs. (They became smaller and smaller as she became bigger and bigger in her hips.) Evelyn E. soon developed unsightly jowls. Her breathing was heavy. She looked flushed upon the slightest exertion. Fearing serious health consequences, she decided she had to do something to bring down her weight. Evelyn E. had always envied many of the movie stars who were so curvaceous and slim, looked so youthful. How could she have a movie star figure?

Enzymes Are Keys To Glamorous Figure. A neighbor who was also a film buff and avidly read as much as she could about the celluloid heroes and heroines, offered Evelyn E. the key to having a movie star figure. The film buff said she read that many of the film stars would use raw fruits to help metabolize their fats, carbohydrates and calories. But most of them literally swore by three special fruits. Evelyn E. bought these three fruits and followed this easy program:

1. Eat these fruits, singly or in combination, whenever you are going to cook in the kitchen and be tempted to taste the foods. Eat these enzyme foods *before* you start cooking, to fortify your digestive system with a supply of enzymes that will be waiting to metabolize foods to reduce cellular expansion.

2. Eat these fruits, singly or in combination, as a raw enzyme salad *before* beginning any major meal.

3. As a beverage, drink their freshly prepared juice as a means of putting a natural control on your appetite and also sending quick-assimilated enzymes throughout your system.

Nibbling Urge Eases, Hunger Ends, Weight Slims Down. Evelyn E. followed this program for 20 days, without any skipping. Soon, she had conquered the urge to nibble and taste-sample while she ate. The gnawing hunger urge ended. More important, her heavy hips slimmed down. Her thick thighs became firm again. She lost some 40 pounds within a few more weeks. She was soon so shapely, she had to buy much smaller sized clothes. She now felt youthfully slim again. She praised the enzymes in the three fruits, available at any corner market, when her husband said she had a lovely movie star figure.

Here are these three fruits and how their Enzyme-Catalyst power can help you realize your objective of being slim as a movie star:

#1—Papaya

This fruit grows on a giant herbaceous plant, and not on a tree. It is a prime source of the enzyme *papain* which attacks clumps of fats in the cell nuclei and helps metabolize them and prevent against fat buildup. *How To Eat:* Freshly sliced with a sprinkle of honey for juicier good taste. Also good as a juice. *When To Eat:* Before you know you will be faced with too many tempting foods. Good as a dessert, too.

#2—Peaches

This fruit contains a small amount of hydro-cyanic acid and fruit ethers. Both of these substances are used as "ammunition" by its enzymes to attack the thick carbohydrate molecules that have clogged up the thick membranes of the cell. Enzymes from this fruit will then help break down these molecules and "slim" down the cells, thereby helping to slim down the entire body. *How To Eat:* Raw, with a sprinkle of honey. Also use sliced raw peaches atop any carbohydrate meal as a natural enzyme-catalyst upon starches to protect against cellular overloading. *When To Eat:* Before any carbohydrate meal. As a replacement for snacks in the kitchen. Also good as a healthful juice. Drink one or two glasses of peach juice before a meal and you will be satisfied with smaller and less-fattening portions.

#3—Grapefruit

Enzymes in the grapefruit alert the metabolism to create a *saprophytic* action wherein calories as well as fats and carbohydrates are "burned up" and washed right out of the cell. Enzymes promote this *saprophytic* action very speedily as they cleanse cells. Grapefruit enzymes help oxidize stored fat from the cells and wash it right out of the system. Stored up or "stubborn calories" that cling to the cells will find it impossible to resist the enzyme's *saprophytic* action and calories can then be speedily metabolized. The submicroscopic cell nucleus in which calories cling can be "washed" with grapefruit enzymes and help keep you permanently slim. *How To Eat:* Peel a grapefruit and eat the sections as a pre-meal enzyme fortification. Use grapefruit wedges with seasonal fruit slices *before* a high-calorie food to give your body needed enzymes for better cellular metabolism. *When To Eat:* As a snack or as a pre-meal food. Also good as a meal in itself with several scoops of skim milk cottage cheese and assorted cheese slices with melba toast crackers. Helps trigger off a chain reaction to promote better cellular metabolism and weight loss.

A VARIETY OF FRUIT ENZYMES FOR FAST, PERMANENT WEIGHT LOSS

Healthy fruit is a prime source of healthy enzymes. When buying fruit, consider those that are in season. Avoid those that are spoiled. Usually higher enzyme containing fresh fruits are free or practically free from blemishes. The healthier the fruit, the more potent the enzymes that creates a catalyst action for fast, permanent weight loss. Here are some popular fruits available at almost any local market:

Apples. They should be firm with good color and flavor. Overripe apples yield to slight pressure and the flesh is often soft and mealy and lacking in enzymes. Select color-rich apples that will have good enzyme content.

Apricots. Highest enzyme content will be found in tree-ripened apricots, if currently available. These are found usually in markets adjacent to the district in which they are grown. High enzyme apricots are plump, fairly firm and a uniformly golden-yellowish color. The flesh is juicy. There should be no signs of decay.

Avocados. Heavy, medium-sized avocados which have a bright fresh appearance and which are fairly firm or are just beginning to soften, usually are highest in enzymes. Pass up avocados that show decay; this can be detected by dark sunken spots.

Bananas. Enzyme content is indicated by the color of the banana skin. It reaches its best quality and flavor after being harvested in the green state. Keep at a warm room temperature for good enzyme ripening. Avoid bruised fruit or those which have dark or blackened areas.

Berries. These include blackberries, boysenberries, loganberries, raspberries which are strong enzyme fruits. Select those berries with bright, clean, fresh appearance. Avoid overripe berries with poor enzyme content. These are dull in color, soft and sometimes leaky.

Blueberries. Super-enzyme power will be found in blueberries that are plump, with a deep, full color throughout the lot. Avoid any decayed berries.

Cherries. Sweet cherries have strong enzyme power. Sour cherries are usually unpalatable but very strong in ability to melt down fats, carbohydrates and calories. A plate of sour cherries can be sweetened with some honey or pure maple syrup. Add cottage cheese. Helps melt enzymes while keeping you satisfied.

Cranberries. Select large bright-red cranberries for potent enzyme activity. Avoid cranberries that show moisture or any injury. This exposure to air and light depletes enzyme supply.

Figs. A ripe fig should be fairly soft or soft to the touch. This suggests good enzyme content. Note that ripe figs sour and begin to spoil quickly. A characteristic odor is noticeable. These should be passed up. Figs are prime sources of carbohydrate-melting enzymes.

Grapes. These should have good color. The skin should be unbroken. Bunches of grapes that have dry and brittle stems and show injured fruit will have lower enzyme content.

Lemons. Those with a fine-textured skin and heavy for their size contain strong enzymes that help metabolize fats. Avoid lemons that have shriveled or hard-skinned appearance. These are spoiled and enzyme supply is low.

Limes. The fruits that are green in color and heavy for their size are highest in enzyme. Deep yellow-colored limes have weak enzyme content. Select firm, heavy limes.

Oranges. Enzymes are potent in oranges that are firm, heavy, with a fine-textured skin and well-colored. Pass up those that have scars, scratches or discolorations as enzymes have "leaked" and potency is weak.

Pears. Pears that are firm or fairly firm, free from blemish and clean, will have good enzyme content. Pears that are soft or that yield too quickly to pressure at the base of the stem are usually enzyme mature and should be eaten quickly. Do not store as they become too ripe and enzymes are evaporated.

Pineapples. A prime source of enzymes. A high-enzyme pineapple has a fresh, clean appearance; it has a distinctive dark, orange-yellow color. The "eyes" are flat and almost hollow. Usually, the heavier the fruit in proportion to its size, the higher the quality of enzymes.

Pineapple enzymes may lose potency if held in a too-dry atmosphere. Keep in a cool room, or refrigerate until ready for use.

Plums and Prunes. A prune is a variety of plum which is treated to a drying process. These are good sources of mild enzymes. Select those that are plump, clean, have a fresh appearance, are full colored, and soft enough to yield to slight pressure.

Quinces. These are strong in enzyme power. Select those that are firm to hard, free from blemish and of a greenish-yellow or golden-yellow color. Bruised quinces show dark discolorations which indicates enzyme loss, so avoid these fruits.

Strawberries. Very good enzyme power. Select those that are fresh, clean and bright, have a solid red color. Avoid those that look dull or decayed since enzymes have been depleted. Strawberries without caps should be passed up since enzymes have been lost through this breakage.

Fresh, raw fruits are prime sources of enzymes which create a "stay slim forever" reaction by metabolizing fats, carbohydrates and calories. They help create the catalyst metabolism that enters into the cells of the body to wash out substances that would cause weight buildup. They are the tasty, delicious ways to keep yourself slim . . . permanently!

In a Nutshell:

1. Steve McK. trimmed his waistline to a new 34 inches, lost 44 pounds, by changing from canned to fresh raw fruits.

2. Raw fruits contain enzymes that "scrub" body cells and "burn up" excessive fat accumulation to help you "stay slim forever."

3. Stella D. was developing secretarial "spread" of her hips. An adjustment called for beginning each meal with raw fruit, ending it with raw fruit. She slimmed down again and lost her "secretary's bottom."

4. Raw fruit enzymes help melt fats, carbohydrates and calories.

5. Follow official U.S. Government suggestions on how to use enzyme fruits.

6. Evelyn E. slimmed down to a "movie star figure" by using three everyday fruits.

7. Enjoy a variety of fruit enzymes for fast, permanent weight loss.

8

How to Use Vegetable Carbo-Zymes For Fast Weight Loss

Fresh, raw vegetables are Nature's powerhouses of more than just enzymes. These are combinations of carbohydrates and enzymes or "carbo-zymes." When you eat a raw vegetable, the energy-building power of the carbohydrates trigger off a dynamic enzymatic action to metabolize fats and calories as well as other carbohydrates. Raw vegetables may well be considered Nature's most potent reducing aids . . . and all-natural, too.

FIVE WEIGHT-LOSING BENEFITS OF VEGETABLE CARBO-ZYMES

A raw, uncooked vegetable, thoroughly chewed and swallowed, then assimilated, will create this set of carbo-zyme slimming actions:

1. *Chewing Helps Starch Metabolism.* When you chew a raw vegetable, you mix it with your saliva, containing the ptyalin enzyme. This creates an alkaline reaction which converts the starch or carbohydrate into sugar. This helps cut down on carbohydrate buildup, and the sugar that is made is in a form that can be better metabolized. Chewing is the first step in slimming, because of this carbo-zyme action.

2. *Digestion Aids In Better Assimilation.* When you eat a heavy fat food, it often feels like a "lump" in your stomach. But if you precede the fat food with a raw vegetable, then your digestive system has a supply of hydrochloric acid, pepsin and renin. These enzymes then transform fatty substances from protein into peptones and coagulate both casein and albumen, two other substances that might otherwise be transformed into fat and stored in tissues. This digestive metabolism, sparked by carbo-zymes, helps control weight buildup.

3. *Better Sugar Metabolism And Oxidation.* Vegetables introduce carbo-zymes into the digestion system so they can convert sugar into dextrose and in a form that is oxidated and assimilated. This protects against sugar buildup in the cells. Carbo-zymes initiate this action so that dextrose can be "burned up" and weight can be controlled.

4. *Controls Fat Accumulation.* The quick energy action of carbo-zymes convert meat and fat-causing proteins into peptones; carbo-zymes then disintegrate fats into glycerine by using the pancreatic enzymes such as trypsin, amylopsin. This method burns up fats, prevents an excess from being stored in the cells. This catalyst reaction is sparked by eating raw vegetables *before* and *after* a particularly heavy meal.

5. *Promotes Better Assimilation.* Vegetable carbo-zymes then use bile (a liver secretion) to emulsify fats, to boost assimilation, to help guard against build-up and storage in the cells where they may cause weight gain.

By eating fresh, raw vegetables, you give your body the needed carbo-zymes that create this catalyst action to give you fast, permanent weight control ... and weight loss.

YOUR WEIGHT-LOSING CARBO-ZYME CHART

Want to lose a lot of weight? Want to lose a little weight? Want to keep your currently slim shape? You can do this through the use of carbo-zymes found in raw or slightly steamed vegetables.

What Are Vegetables? Basically, they are plants cultivated for their edible portions. This definition includes leaves, stems, roots and tubers, pods, buds, flowers, seeds and fruits. They are prime sources of weight-losing or weight-controlling carbo-zymes.

This chart shows you which vegetables contain those carbo-zymes that can help you reach your desired weight goal.

1. *Eat Vegetable Leaves For Fast Weight Loss:* spinach, chard, beet greens, turnip greens, chinese cabbage, mustard greens, Brussels

sprouts, cabbage, kale. **ECD Action:** These carbo-zymes are powerful with activity and can help burn up unwanted fats in the cells.

2. *Eat Stems For Slow-Steady Weight Loss:* Rhubarb stalks, celery, cardoon, fennel, asparagus, carrots, salsify. **ECD Action:** These carbo-zymes are more liquid and will offer time-release action in melting away unwanted carbohydrates.

3. *Eat Roots And Tubers For Modest Weight Loss:* Turnips, radishes, parsnips, Jerusalem artichokes, onions, potatoes. **ECD Action:** These carbo-zymes have been developed underground, out of the sun's action, and have a mild catalyst action in slow and modest weight loss.

4. *Eat Buds And Flowers For Calorie Control:* French artichoke, broccoli, cauliflower, beets, cucumbers, squash. **ECD Action:** These carbo-zymes act very specifically upon calories and exert an oxidating effect to help metabolize them to protect against cellular weight.

5. *Eat Seeds For Better Weight Control:* beans, peas, peanuts, okra, most seeds, which are considered part of the vegetable family. **ECD Action:** These are living foods, created by the plant in the reproductive part of the vegetable and are teeming with valuable carbo-zymes that help control weight.

6. *Eat "Fruits" Of Vegetable Plant For Keeping Current Slim Shape:* tomatoes, green peppers, pumpkin, eggplant, garlic, mushrooms, corn. **ECD Action:** Called "fruits" of the vegetable plant, they are the end product of the growth. They contain mild but vigorous carbo-zymes that keep your metabolism alert and working efficiently so that you can keep slim.

Above-Ground Vs. Under-Ground Vegetables. Those grown above-ground are sun-nourished and their carbo-zyme power is considered to be stronger. Those vegetables grown under-ground are often considered of secondary (but potent) carbo-zyme power since the sun has not nourished them from the moment of their "birth." A good Enzyme-Catalyst Diet Program will feature both types of vegetables with emphasis upon above-ground vegetables for faster and more permanent weight-loss.

THE MIRACLE WEIGHT-LOSING, HEALTH-BUILDING POWER OF VEGETABLE CARBO-ZYMES

Many diets cause a feeling of weakness and ill health, while weight is lost. The difference with the Enzyme-Catalyst Diet is that it offers you tasty food and *lets food slim you down!* Furthermore, the food will build your health so you can enjoy dieting!

From Diet Agony To Diet Pleasure In Three Weeks. Vera Q. had tried one diet after another with mixed results. While she may have lost weight, she lost her health with the pounds. Most deprivation and starvation diets made her nervous, anemic, susceptible to colds, so weak that she could barely drag herself through the day. She wanted to *eat and reduce.* A fellow food-lover told her that the Enzyme-Catalyst Diet could help her lose weight, while building her health! She had to follow this basic rule:

1. Begin each meal with a large raw vegetable salad. Use a salad dressing of equal parts of apple cider vinegar with polyunsaturated oil. Chew thoroughly.

2. Cooked vegetables were permitted. They should be steamed in as little water as possible for as brief a time as possible. This helped preserve much of the carbo-zyme potency.

3. Gradually, replace excessive amounts of heavy foods, with more raw vegetables.

Slims Down, Feels Healthy. It was this program that made it a joy for Vera Q. to lose up to 36 pounds within 11 weeks. But more important, she felt fit as a fiddle. She was not weak. She did not feel nervous. She later exclaimed that dieting on the ECD Program was so delicious, she would stay with it . . . and keep permanently slim, too.

HOW THE ENZYME-CATALYST DIET BUILDS HEALTH

Carbo-zymes in the raw and slightly steamed vegetables promoted metabolism so that unwanted fats and calories could be dissolved out of the cells. But more important, the carbo-zyme catalyst action promoted the formation of red blood cells, stimulated respiration and nitrogen metabolism of the cell tissues. This created better metabolism of protein, normalized blood pressure, corrected pancreatic function, improved circulation which added up to better health while slimming down.

Unique Catalyst Benefit: Thoroughly chewed raw vegetables, created a unique health-building and weight-losing reaction. Enzymes restored a favorable acid-alkaline balance of the ashes after combustion of foods. They supplied a protein of very high biological value that caused improved cellular repair but with reduced caloric buildup. This enzyme catalyst reaction within the body is the key to better health and then creating fast but permanent weight loss through cellular slimming and nourishing.

A VARIETY OF VEGETABLE CARBO-ZYMES FOR FAST, PERMANENT WEIGHT LOSS

Raw vegetables offer carbo-zymes that alert the metabolism to help cleanse the heavy, weighty molecules that cling to the adipose cell tissues. With this catalyst action, the adipose cells are cleansed of their stored up fats and calories. Weight loss is fast and permanent when there is a daily supply of vegetable carbo-zymes, available through fresh raw vegetables. Here are some of the high carbo-zyme vegetables available at any local market:

Asparagus. Enzymes will use the alkaloid known as asparagine to break up accumulated fats in the cells and alert the metabolism to wash weight out of the system. Use slightly steamed asparagus for good enzyme benefits.

Beets. Mix a little of shredded beets with a raw salad. Beets are able to penetrate the red blood corpuscles to promote a fat cleansing reaction. Its enzymes will use potassium and vitamins to maintain healthy cells and guard against fat buildup.

Carrots. Enzymes take the Vitamin A precursor and use it to activate metabolism and promote better cellular metabolism to ease fat buildup and bring down excess weight.

Brussels Sprouts. Slightly steamed or shredded, this vegetable has a supply of highly concentrated enzymes that help cleanse accumulated wastes from the cellular system. Combine with a fresh fruit juice dressing for added power.

Cabbage. Enzymes will break up wastes and then wash them out of the system. Cabbage enzymes will also use its high sulphur and iodine content to regulate glandular secretions so that metabolism is alerted to promote better cellular cleansing.

Celery. A supply of natural sodium which is used by enzymes to slough off calories and carbohydrates from the walls of the cells. Enzymes will use celery's natural sodium to dislodge fats and then transport them to the liquid portion of the body for elimination.

Cucumber. Enzymes are especially vigorous in cucumbers; they take its high silicon and sulphur content to help wash out uric acid as well as other wastes from the system. Enzymes will also use potassium in the cucumber to help metabolize fats, carbohydrates and calories for elimination.

Endive. This is a curly vegetable, also known as chicory or escarole. It resembles lettuce. Its enzymes use its essential minerals to dislodge accumulated wastes in the system to help remove them from the cells.

Lettuce. Enzymes use its rich supply of iron and magnesium to help alert a better circulation so that metabolism can dispose of accumulated weighty substances. Enzymes will use magnesium from the lettuce to maintain better blood fluidity so that fats, carbohydrates and calories can be transported more easily for removal from the body.

Parsley. Enzymes in this herb plant stimulate oxygen metabolism which alerts the adrenal and thyroid glands to promote a catalyst action on body capillaries and arterioles so that weight is washed off.

Parsnip. This vegetable has strong root enzymes which use its rich silicon and sulphur content to work with its potassium supply to boost a sluggish metabolism and help catalyze accumulations from cells. This helps maintain good body balance and better weight loss.

Pepper (Green). Enzymes from this green plant will use minerals to boost better blood metabolism of fats, carbohydrates and calories.

Radish. Combine with carrots for stronger enzyme action. The radish enzymes help restore tone of the various mucous membranes of the body, helping to cleanse the body of sludge which contributes to weight gain.

Tomato. Here is a good souce of alkaline substances needed by enzymes to help metabolize accumulated sugars and starches. Enzymes will use the tomato's very high citric and malic acids to boost the process of metabolism which will then help control appetite and weight.

Turnip. Enzymes are potent in the turnip. Enzymes will use this vegetable's calcium and phosphorus to create an alkaline reaction in the system and help reduce hyperacidity. This creates a more favorable environment for cleansing the cells of accumulated wastes.

Watercress. Here is a plant that has a high supply of enzymes which are used to cleanse the intestinal tract and guard against accumulated guars and starches. Enzymes in watercress use sulphur, phosphorus and chlorine to promote regeneration of the blood which then increases oxygen transmission through the tissues and cells. This action dislodges accumulated carbohydrates and calories and paves the way for better and more permanent weight loss.

Fresh, raw vegetables, eaten as a salad or even as a meal with a healthful dressing of equal portions of apple cider vinegar and oil, will help give your body the working materials (enzymes) for fast, permanent weight loss through cellular metabolism.

HOW CARBO-ZYMES MELTED "STUBBORN WEIGHT" AROUND WAISTLINE

Paul J. had tried one diet after another with temporary results. If he did lose weight, it would be gained right back again as soon as he went off the particular diet. He liked to eat and to deny himself his favorite foods was tortuous. Paul J. may have lost much excess weight, but he could not get rid of a "spare tire" of stubborn weight around his waistline. It seemed to increase in girth, as other body parts remained the same. Paul J. wanted to get rid of this "stubborn weight" in the worst way but had given up on many of the diet plans that gave him "see saw" weight fluctuations.

Discovers Power Of Raw Carbo-Zymes. He vacationed in an island area where heavy meat foods were so expensive because of shipping difficulties, that he began to live on fresh raw vegetables, as well as fruits and whole grains. Paul J. discovered that three days of such a healthy program started to shrink his waistline. He began to pull in his belt. Before the week was over, he had taken in his belt so much that he counted *four inches lost from his waistline.* He also took in some *six notches* in his belt. His "spare tire" evaporated. His buttocks slimmed down. Within ten days, his "stubborn weight" had surrendered to the fat-melting powers of carbo-zymes. When he returned home, he changed his eating habits. He would now begin each meal with a raw vegetable salad. For in-between snacks, he would munch on raw vegetables.

He could continue eating his favorite lean meats, poultry, eggs, dairy foods. But *before* any such meal, he needed to fortify his system with fat-melting enzymes from raw vegetables. Now, Paul J. had a slim 34 inch waist. He no longer was troubled with "see saw" or "up-and-down" weight fluctuations. Carbo-zymes took his weight off fast and permanently!

HOW CARBO-ZYMES SLIM DOWN OVERWEIGHT BODY CELLS

As catalysts, enzymes from raw vegetables penetrate the walls of the capillaries into the compartment called the *interstitial space* (or, the space between the cells). Here, the carbo-zymes go through the cell wall, and boost the action of *osmotic pressure*. That is, carbo-zymes use vitamins, minerals and proteins to wash away accumulated molecules. This is often referred to as tissue *turgor,* of the heavy weight that is given to certain spaces when they are occupied with excessive accumulations. Carbo-zymes catalyze these molecular weights and by creating a micro-electronic action in the cell, itself, help slim down overweight.

Daily Vegetables Offer Permanently Slim Cells. A daily supply of raw vegetables will offer your cells an ample amount of carbo-zymes which will help create around-the-clock *osmotic pressure* to keep your cells slim. Carbo-zymes will work while you sleep, too! They are the natural way to give you fast, permanent weight loss.

How Raw Greens Promoted Youthful Slimming. Betty H. loved good food. She said she always ate vegetables as well as fruits, but still was some 35 pounds overweight. To make it more uncomfortable, she kept on gaining more and more weight. She developed a "double chin" as well as heavy arms and heavy thighs. She said she felt that enzymes would do nothing for her. But the fault was not with the enzymes. The fault was with the vegetables Betty H. ate. They were cooked! When she followed a very simple enzyme-catalyst rule, she began to lose weight. Before too long, she not only lost the 35 pounds of overweight, but her chin was slim, her arms and thighs became youthfully slim, too. More important, she stopped gaining weight.

Simple Enzyme-Catalyst Rule: All vegetables that can be eaten raw, should be eaten raw. All vegetables that must be cooked, should be steamed ever so lightly in a little bit of water or oil.

Benefits Of Raw Vegetables: Raw vegetables, *especially greens,* contain bulk or fiber. Enzymes transform these into a highly magnetized form as they pass through the intestines. They draw from the body accumulated fats, carbohydrates and calories, as well as wastes from the tissues. Enzymes use this bulk as a broom and vacuum cleaner on the cells to keep them slim. When the vegetables are cooked, the action is more like that of a mop. Weight still remains. Raw greens are prime sources of bulk which act as "living magnets" in the body to suck up weight-building substances and then cast them out, via elimination. *Enzymes are alive in living vegetables!* They are either sluggish or dead in cooked or dead vegetables! This simple adjustment can help create the catalyst action needed to keep you permanently slim.

EIGHT WAYS TO PREPARE YOUR CARBO-ZYME VEGETABLES

1. *Combinations:* Select contrasting texture, color, form and flavor when selecting raw vegetables for salads. Plan your carbo-zyme salad as part of your meal or as a meal in itself.

2. *Cold, Crisp Ingredients:* To preserve carbo-zyme power, keep vegetables (especially salad greens) cold and crisp after washing, until ready to use. Prepare salad as close to serving time as possible.

3. *Variety:* Combine various salad greens to give interest to salads. Choose dark-green ones more frequently for good carbo-zyme vigor.

4. *Keep Attractive Sizes:* Because enzyme activity begins with sight and salivary stimulation, sizes should be attractive. Slice vegetables so they can be easily chewed. Arrange attractively on the plate.

5. *Toss Ingredients Lightly:* Do not bruise or crush raw vegetables since this breakage causes enzyme evaporation and catalyst weakness. Be gentle. Add juicy tomatoes last. Layer the salad ingredients whenever possible.

6. *Use Enzyme-Catalyst Dressing:* Fresh, raw fruit juice helps boost enzyme power of the raw vegetable. Use a light, delicate dressing on a salad that accompanies a substantial meal. Use a more substantial dressing on a salad that functions as a major meal. Use a minimum of dressing. Too much will "drown" enzymes and reduce their effectiveness. A favorite enzyme-catalyst dressing is a combination of apple cider vinegar and oil with a bit of honey for added taste.

7. *Keep Casual Look:* Stimulate enzymatic flow in your digestive system with a casual salad plate. Keep it simple. It helps alert a better catalyst response that begins with mental anticipation of the delicious salad to be eaten.

8. *Service:* Serve your salads from a large bowl or a deep platter. Arrange your raw vegetables on individual serving plates, or in small salad bowls. This makes it easier and more enjoyable. Since digestive enzymes are so influenced by the emotional state, it is important to make salad eating a joyful time. This boosts catalyst power of the raw vegetables.

14 CARBO-ZYME SALAD DRESSINGS

Boost the power of carbo-zymes with a healthful salad dressing that, itself, is rich in weight-melting enzymes. Here are 14 of these tasty and catalyst-type salad dressings:

1. Apply just a sprinkle of sea salt. (Available at most health stores.) A prime source of ocean minerals that alert the body's enzyme-catalyst system for better metabolism and weight loss.

2. Sprinkle lemon juice or any citrus fruit juice on the salad.

3. Sprinkle on fruit juice, a bit of sea salt, and some vegetable oil.

4. Combine lemon or orange juice, some honey, a little water. Shake vigorously in a bottle for a healthy enzyme dressing.

5. Combine a few grains of sea salt, a squeeze or two of lemon, and a quarter teaspoon of honey.

6. Combine yogurt (fermented food which is brimming in enzymes) with lemon juice.

7. Combine low-fat cottage cheese with any fresh fruit juice.

8. Combine buttermilk with yogurt and fruit juice.

9. Combine fruit juice, lemon juice and honey.

10. Combine cottage cheese, lemon juice, mashed avocado and sea salt.

11. Add finely chopped vegetables which have a distinctive flavor such as radishes or beets. Also try herbs mixed with an oil dressing. Try dill in an oil dressing.

12. Combine chopped hard-cooked egg, olives, pimiento to a low-calorie mayonnaise.

13. Add rolled chopped nut meats or nut butter to low-calorie mayonnaise.

14. Combine grated fruit rind and sprinkle over your salad.

ECD Benefits: These 14 carbo-zyme salad dressings add more than exciting flavor to your raw vegetable salads. They, themselves, are prime sources of very potent enzymes which amplify the catalyst vigor of the enzymes in the raw salads. This combination helps promote better *osmotic pressure* which is the key to catalyzing weighty substances in and through the adipose tissue cells. It offers a tasty and natural way to take off weight fast and permanently! You can, actually, *eat your way thin!* Carbo-zymes make it possible to keep you thin, too!

In Review:

1. Raw vegetables offer five weight-losing benefits through their all-natural carbo-zyme action. Works while you sleep!

2. To keep weight off, follow your weight-losing carbo-zyme chart before eating any fat, carbohydrate or calorie food.

3. Vera Q. gained health and lost some 36 pounds in 11 weeks on the simple enzyme-catalyst program.

4. Eat your way to permanent slimness with a tasty variety of carbo-zyme raw vegetable foods.

5. Paul J. lost stubborn weight, melted his "spare tire" and took in six belt notches on an easy carbo-zyme program. Carbo-zymes create an *osmotic pressure* reaction that slims down overweight body cells.

6. Follow the easy 8-step program to prepare carbo-zyme vegetables.

7. Amplify the catalyst power of enzymes with any of the 14 carbo-zyme salad dressings.

9

How Enzymes
Help De-Control
Your Compulsive
Eating Urges

Are you a compulsive eater? Take this test to determine whether you have an obsessive eating urge or one that is only average. Write "yes" or "no" on a piece of notepaper to these questions. Be honest, for your own benefit.

1. Do you eat even if you are not hungry?

2. Do you go on eating binges for no obvious reason?

3. After overeating, are you filled with reactions of guilt and remorse?

4. Do you spend too much time thinking about food?

5. Do you anticipate a great pleasure for the moments when you can eat alone?

6. Do you plan such secret eating binges ahead of time?

7. Do you eat moderately when with others, but stuff yourself when alone?

8. Is your weight interfering with your daily living?

9. Have you dieted for a week (or more) and still ended up overweight?

10. Do you feel annoyed when others tell you that just "willpower" will control your overeating urge?

11. Do you keep telling yourself that you can diet "whenever you wish"?

12. Do you have an eating urge, other than regular mealtime? During odd daytime and night hours, that is?

13. Do you find an escape from problems or worries with food?

14. Have you ever been medically treated for overweight?

15. Is your eating urge causing you, or others, unhappiness?

Your Score: If you replied "yes" to three or more of these questions, it indicates that you are a compulsive eater. Or else, you are well on the way to becoming one.

Admit Your Weakness. As stated at the outset of this self-test, be honest with yourself. Admit that one bite is leading to another, and another, and another. Admit that compulsive eating is your weakness. Once you have this positive attitude, you can then be helped by the enzyme-catalyst power of everyday foods.

Enzyme-Catalyst Diet Works Where Other Programs May Fall Short. To de-fuse your compulsive eating, the Enzyme-Catalyst Diet works by "turning off" you hypothalamus, the brain segment which triggers off the eating urge. Once this is self-controlled, there is less of a desire to eat. The adipose tissues are spared excessive buildups. Other programs *ask you* to use willpower. This is difficult. Willpower originates in the hypothalamus. You are often unable to control this segment of the brain. An outside source of control is needed.

Enzymes Satisfy Willpower. When your digestive system metabolizes the enzymes from raw foods, they are absorbed into the bloodstream. They are transported throughout the body, especially to your hypothalamus. Here, fructose from fruits and carbo-zymes from vegetables work to soothe the hypothalamus. This offers a satisfaction value to your willpower. Enzymes work to soothe your hypothalamus and the eating urge becomes controlled.

Here are basic causes of overeating and ECD solutions.

Question: Dieting can be very dangerous to one's health. Why risk health by reducing?

ECD Answer: A diet that deprives you of important food can be dangerous. But overweight is dangerous, too. The ECD answer is to use raw foods that have enzymes to satisfy your hypothalamus. They also raise your blood sugar levels so that there is less of a compulsive eating urge. Enjoy a *balanced diet* on the ECD Program. Just remember to eat raw fruits or vegetables (or their juices) before and after each meal.

Question: Dieting is so weakening that it often causes more overeating at the end. What can be done to guard against fatigue while dieting?

ECD Answer: Fatigue can be eased by a high fruit program. Fruits contain levulose which is a speedy source of natural energy, helping to satisfy your glands, controlling your appetite, too. Fruits offer *enzyme energizers* that keep you alert and active while slimming down.

Question: There are times when a person feels sorry for himself (or herself) and uses eating as self-appeasement. How can this be controlled?

ECD Answer: Go ahead and eat when you feel self-pity. But the secret here is to eat low-calorie foods that alert enzymes. Low-calorie crispy raw vegetables such as celery, carrots, lettuce, cucumbers, raw cabbage, assorted seeds and anything "chewy" in the plant world, will alert your digestive juices. This offers solace, without filling up.

Question: If there are recurring cravings for cookies, snacks, candies, can the ECD Program help control weight gain?

ECD Answer: If you must eat sugary and calorie-high confections, then "sandwich" them between raw plant foods. That is, eat a few raw radishes, celery wedges, shredded carrots *before* the sweets. Eat a few more raw plant foods *after* the sweets. This offers your digestive system double-barrelled protection against weight buildup since this supply of enzymes will metabolize the sweets and protect against cellular clogging. But try to limit amounts of confections eaten. Raw plant foods will help you do this.

Question: A problem with most reducing diets is "hunger headache." Will this be the same problem on the ECD Program?

ECD Answer: Such a problem plagued Gene I. She always felt head pains, throbbing temples, gnawing aches throughout her head and shoulders, when on a reducing program. She knew she could lose weight if only she could avoid those "hunger headaches" and related pains. She followed the ECD Program but with one added feature.

Whenever she felt the onset of any "hunger headache," she would drink a glass of fresh fruit juice. **ECD Benefit:** Enzymes took up the fruit's levulose which it used to raise the blood sugar levels and thereby controlled the hypoglycemia (low blood sugar) type of headache. Gene I. lost weight, and kept it off, without any "hunger headaches" with the use of fruit juice enzymes.

Question: Most diets can take off pounds . . . but they go right back on after a cruise vacation where lots of eating is part of the pleasure; also, many long weekends or festivities add pounds. How can the ECD program control this compulsive or unavoidable eating urge?

ECD Answer: Enzymes need a high-protein intake to offer satiety value that will then control the urge to take of excess foods. Enzymes use protein to "burn" a "steady flame" of satisfaction that controls the eating urge. A suggestion is to eat a hard-boiled egg, or a portion of skim milk cottage cheese, or a chicken drumstick, or some lean fish *together* with a raw fresh fruit. The plant enzymes will then take the meat protein to help give you satisfaction so you do not overeat. Most "weekend eating binges" are caused by a deficiency of enzymes and proteins. This creates "hollow hunger" and an urge to eat confections. Fortify your body with some of these protein foods and enzymes help reduce the weekend or vacation eating urge . . . and reduce weight, too.

Question: Business often obligates people to eat heavily during lunchtime. How can this be controlled? The food is often irresistible.

ECD Answer: Tell your business luncheon guests that you are on a special slimming program. Then ask for *half portions* of all foods. Most important, begin with a fruit salad. End with a platter of fresh raw fruits. The enzymes will work to help metabolize even those fats, carbohydrates and calories you have eaten in smaller amounts. If you have a particularly big lunch, then cut down on your evening meal.

Question: My wife keeps putting more and more food on my plate, even though she knows I am overweight. How can I eat less without offending her?

ECD Answer: Begin by asking for wedges of raw vegetables. These offer enzymes that will soothe your hypothalamus so your compulsive eating temptation will be controlled. Then when you are in possession of your willpower, eat only half portions. Leave the food over on the plate and say you are stuffed. Do this often enough and you'll soon be getting smaller portions.

Question: Whenever I diet, I become mean and unpleasant. I snap at others. How can this be avoided on the ECD program?

ECD Answer: Diets that deprive you of foods, can make you irritable. You want to satisfy your stomach, simply speaking. Do this with luscious sweet fruits at the start and ending of your meals. Enzymes will raise your blood sugar levels and also wash your adipose cells so there is an internal balance that makes you feel more contented.

Question: Much weight is traced to processed meats, pastrami, corned beef and the like. I enjoy these meats. That's why I keep gaining weight. How can I de-fuse my urge for these types of meats?

ECD Answer: Begin with a plate of a *fermented* food such as homemade sauerkraut, dill pickles or yogurt. These foods contain *lactic acid ferments* which are used by enzymes to help empty the adipose cells of clinging fats, carbohydrates and calories. Keep the body fortified with these *lactic acid ferments,* since they are "scrubbers" needed by enzymes to protect against weight buildup traced to heavy, marbled meats and processed meats, too. If a meal is begun with an ample amount of these low-calorie but high-fermented foods, there is a control put on the appetite, too.

Question: Boredom and frustration often causes overeating. How can these problems be eased by the ECD program?

ECD Answer: Relieve boredom by enjoying luscious fruits and vegetables. Frustration can often be dispelled by preparing elaborate raw fruit and vegetable salads with a variety of different enzyme salad dressings. Just preparing such foods will help ease your emotional boredom and offer satisfaction. Eating these raw foods will satisfy your appetite without adding weight.

Question: Because of ulcer-like stomach pains, dieting was difficult so I would drink lots of milk every day. This added weight. How can your catalyst program protect against stomach growling?

ECD Answer: Pamper your ulcer-like stomach reactions by drinking *skim milk* which is low in calories. Then indulge in bananas after the milk is finished. Bananas are low in fat and have *no* cholesterol. The enzymes in the bananas can work freely as they take up the protein in the milk and "coat" the stomach so there is a feeling of contentment. Even when eating less, you should feel comfortable with this combination of skim milk and high-enzyme bananas.

Question: My habit is to eat a thick wedge of pie and a glass or two of milk at bedtime. Otherwise, I can't fall asleep. This is putting on extra pounds. But I can't control this habit. What can I do?

ECD Answer: This problem caused Louise O'C. to put on some 20 overweight pounds which she could not take off. But without her pie and milk at bedtime, she suffered such insomnia, she would get up in the middle of the night and raid the refrigerator. This added still more pounds. To de-fuse this eating habit, she followed an easy and delicious enzyme-catalyst principle: satisfy the hypothalamus and the habit-forming segment of the brain with enzymes. Replace the pie with a bowl of seasonal berries, flavored with a bit of honey. Use skim milk flavored with fresh berry juice. Enzymes in the fruit take the levulose from the honey, combine with the protein from the milk and bathe the adipose tissues, as well as the cells of the hypothalamus with soothing comfort. This offers satisfaction and de-fusing of the appetite. Louise O'C. found this simple change satisfied her taste and helped her sleep the night through. She soon lost 20 pounds . . . and an additional unwanted 12 pounds soon melted away.

Question: Diets take off my weight but a slight change or delay only puts the weight back on. How can this be corrected so that the weight goes off and stays off?

ECD Answer: This is the "see-saw" or "Yo-Yo" syndrome. Weight goes off, but it often comes back. Under the Enzyme-Catalyst Diet program, this problem is eased by using enzymes to keep your metabolism working efficiently. You need to have thoroughly chewed fruits and vegetables daily. Fit them into your menu plan. Use them daily. Use for desserts, even as meals, in themselves. They are tasty, effective. Enzymes will help satisfy your stomach contractions, but more important, they alert your metabolism to keep functioning without any delays. This helps burn up fats, carbohydrates and calories, even if you tend to overeat on occasion or go off any reducing program. This does not mean that you can overeat. Rather, you fortify your body with fruit, vegetable enzymes so they can be used by your metabolism to guard against adipose tissue or cell flattening. But your eating urge is eased, too, by the hypothalamus-satisfying enzyme reaction.

TEN ENZYME-CATALYST DIET TIPS TO CONTROL YOUR APPETITE

Retrain your eating habits. Help turn off the signal to eat, with these Enzyme-Catalyst Diet Tips:

1. Be sure to eat a good breakfast, one that is high in protein as well as enzymes. This is important since enzymes will burn protein slowly and help de-fuse your compulsive eating urges throughout the day.

2. When you eat, take your time. Enzymes need this time to help raise your blood sugar. This also helps to make smaller amounts of food more satisfying.

3. Replace a "coffee break" with a raw juice break. A vegetable juice is satisfying and eases the appetite urge, too.

4. If you get hungry even though you've just eaten, try a glass of tomato juice, sprinkled with lemon juice. Or munch on celery, carrots, radishes, pickles. These offer enzymes to help control appetite.

5. Chew your food thoroughly. Enzymes in the saliva are alerted to the responsible tasks ahead. This chewing process also alerts digestive enzymes that will better metabolize and burn up fats, carbohydrates and calories.

6. Think and concentrate on your food. Give it your undivided attention. This emotional rapport creates enzyme harmony and satisfaction. Many folks talk or watch television while eating. They keep on eating and eating, without proper chewing. They also wash half-chewed food down with liquids. This causes drowning of enzymes and fat buildup. Chew foods carefully and let your mouth liquids wash down the food. Drink liquids no sooner than two hours after a meal so your digestive enzymes can work without interference.

7. Protein is important in your daily program. Combine a protein food with an enzyme food. This helps raise your metabolism, burn up stored fat, eases hunger pangs. A protein-plus-enzyme meal such as meat with fruit, fish with vegetables, dairy with fruit, eggs with fruit or vegetables, boosts your metabolism so that combustion of fat for energy is increased. Enzymes thus help you lose weight fast . . . and permanently. Protein is used by enzymes for this metabolic process to keep weight off, and control your appetite, too.

8. Whenever you eat, sit down! If you eat while standing, or eat on the run, your enzymes cannot function efficiently. Metabolism goes awry. There is an urge to keep on eating. So sit down for a meal. It creates better digestive power and enzymes can work without feeling "tense" or nervous.

9. If you have a compulsive urge to eat, *then schedule one*

time each day to surrender to that impulse. Plan ahead for that specific time. Then when you eat your "forbidden" indulgence, do so very slowly and extract full enjoyment. This will help ease your guilt feelings about "cheating" on your reducing program. But more important, it eases your eating urge and can also help ease the urge to continue "cheating." When you condition yourself in advance to enjoy the "forbidden" food, your enzymes are receptive to better metabolism. Mentally, you will soon feel less of a desire to "cheat" when the "fun" is taken out of it by pre-scheduling. This "cheating" is allowed only at this written-down time. Not before. Not after.

10. Use easy self-hypnosis in weight control. As you partake of a heavy-calorie food, or any excessive amount of food that will put on pounds, ask yourself, "Is this portion necessary? Must I finish every little bit? Can't I put much of it away for another time? Would it be so tragic if I stopped eating now?" Then tell yourself, "I'll lose unwanted flab, I'll tighten up my stomach, I'll firm up my buttocks, I'll slim down my thighs, I'll look youthful, if I'll just cut down my portions." Conjure a vision of a new "slim you" and the urge to eat should begin to subside.

By being honest to yourself, by using wholesome foods, by including raw enzyme foods with each and every meal (or as meals, in themselves) you should be able to de-control your compulsive eating urges.

Furthermore, enzymes will catalyze accumulated fats, carbohydrates and calories and help slough them out of your adipose tissue cells. As weight goes down . . . so does your appetite, thanks to the power of the Enzyme-Catalyst Diet Program.

Important Facts:

1. To determine if you are a compulsive eater, take the easy test.

2. Weight caused by poor willpower, inability to stick to a diet, and problems of daily living can be corrected by using the ECD program suggestions.

3. Gene I. overcame "hunger headache" with a fresh fruit juice drink, brimming with enzymes that soothed her blood sugar levels.

4. Before eating heavy meats, enzyme-activate your system with a plate of any fermented foods.

5. Louise O'C. used a few enzyme foods as a "nightcap" to break her "pie and milk" weight-gaining habit. She enjoyed good food while she lost more than 20 excess pounds.

6. De-control your appetite with 10 helpful Enzyme-Catalyst Diet tips.

25 Natural Enzyme Foods That Act as Organic Reducing Pills

Reducing pills cannot melt pounds away. Instead, they create a state of physical or emotional dependence. They addict the overweight person. They create a feeling of euphoria and a lessened desire for food, but not without serious side effects. Reducing or diet pills have their hazards. They can be habit-forming. They can often be fatal. Let us look at diet pills more closely.

Amphetamines Or "Pep Pills." These drugs are stimulants. They act on the nervous system and give the dieter a "lift" or an improved sense of well-being. They are used medically in the treatment of mild degrees of mental depression and in certain ailments of the nervous system. They depress portions of the central-nervous system and help ease the urge to eat. **Health Risk:** Amphetamines can lead to a psychic dependence that can have serious consequences. The dieter may develop a tolerance that causes increasing amounts of the drug to be taken. A *single large overdose* may cause severe anxiety, convulsions, cardiovascular and gastrointestinal disturbaces. Excessive amounts may cause tremor, intensive restlessness, insomnia, mental confusion, constipation, headache and panic reactions.

Even in prescribed amounts, amphetamines can raise blood pressure or adversely affect circulation. This can be risky for the overweight who has heart trouble or circulatory disorders.

Barbiturates or Sedatives. These drugs are often prescribed for the person who is too restless to sleep and may be tempted to raid the refrigerator at night. Or for the person who has a nervous compulsion to eat. This so-called sedative can cause drowsy reactions, lethargy, even emotional depression. **Health Risk:** A dieter with a cardiovascular disorder may further disturb his health with the use of a depressant drug. Chronic and excessive use can lead to physical addiction as serious as that caused by narcotics. A large overdose can be fatal. They also interfere with the action of the central nervous system.

Diuretics or "Water Pills." These drugs help eliminate water from the system. They deplete body salt and increase the flow of urine. They are often prescribed to relieve conditions such as edema (swelling) caused by congestive heart failure. **Health Risk:** There are serious side effects of weakness and tiredness. Excessive water loss means the washing out of essential vitamins, minerals, enzymes. The body may become malnourished. Furthermore, diuretics may cause weight loss but this is meaningless because *the loss is that of fluid and not fat.* The fluid is soon replenished and weight is regained.

Hormone Drugs. Various hormones such as thyroid and *human chorionic gonadotropin* or HCG may cause weight loss by correcting any glandular deficiency, if that is responsible for overweight. **Health Risk:** These hormone drugs may produce serious side effects, including cardiac disturbances. They also cause the loss of muscle tissue, calcium and nitrogen from the body. This may cause serious muscular weakness or excessive bone destruction. When hormone drugs are taken in sequence, with other drugs, there may be an interaction of their chemical components and lead to serious illnesses, even fatality!

Appetite Suppressors. These products usually contain a drug called *phenylpropanolamine.* It reportedly will lessen the appetite and control compulsive eating. **Health Risk:** Larger amounts of this drug are needed to control the appetite and such amounts can only be obtained by a physician's prescription. In an undetected cardiac condition or circulatory problem, this type of drug can cause serious repercussions.

Bulking Agents. These products usually contain non digestible, harmless substances (usually *methylcellulose*) which swell up in the body and give a feeling of satisfaction. This is supposed to reduce the appetite. **Health Risk:** These bulking agents do not really satisfy the desire for food and the urge to chew and eat. Furthermore, the *methylcellulose* may often bulk up in the digestive system and remain lodged. In the absence of natural bulk agents such as whole grains, leafy vegetables and fruits, this synthetic bulking agent may cause obstructions in the digestive system.

Over-The-Counter Reducing Pills And Aids. There are hundreds of non-prescription diet products available in supermarkets, drug stores, through magazine, newspaper, television and radio advertising. The claim is that these products can cut down on your appetite, "flush" your weight away or reduce your blood sugar levels. **Health Risks:** These products contain chemicals and additives and mixtures of medications that may cause adverse side effects. Since they are available without prescription, the temptation is to use them in large doses. This can be serious to your health. Medical authorities say these over-the-counter have little or no real medical effect in helping people reduce. Even if there is weight loss, it is usually of a temporary value. The safety and effectiveness of the ingredients used in reducing aids and weight control products are always under question.

HOW RAW FOOD ENZYMES ACT AS ORGANIC REDUCING PILLS

Enzymes in raw foods, such as fruits, vegetables, seeds, nuts, grains, fermented foods, act by stabilizing the metabolism so that there is a control on the secretory processes. This creates an inhibitory effect so that there is a normal desire for food. Raw food enzymes soothe the hunger urge, control the gastric secretions so there is a lessened desire to overeat, promote a feeling of satiety which puts a natural stop-gap on the urge to overeat.

Create Natural Slimming Action. Basically, overweight is traced to an accumulation of fats, carbohydrates and calories that are lodged as thick molecules in the adipose cell tissues. Enzymes work to dislodge these heavy accumulations and metabolize them so they can be eliminated from the cells and washed out of the body. This process is more than a natural slimming action. It stabilizes the metabolism and controls the eating urge by creating a "slow burning" of satisfaction. The three benefits of raw food enzymes are:

1. *Fat-Splitting.* This is a lipolytic action in which enzymes pierce

the clumps of fat in the adipose cells and help burn them up through metabolism. A healthy metabolism is the body's self-regulatory mechanism for a healthy and normal appetite.

2. *Carbohydrate-Splitting.* This is an amlolytic action in which enzymes break up accumulated carbohydrates in the cells, and split them so they can be better disposed of through metabolism. Steady carbohydrate metabolism also offers an emotional satisfaction and reduced eating urge.

3. *Calorie-Melting.* This is a calorific-lytic action in which enzymes help dissolve calories that have become excessively stored in the adipose cell tissues. They further promote a feeling of satisfaction because of their calorie burning action. This helps ease the urge to eat since calories are being utilized by enzymes and metabolism is at a healthy level.

Raw Foods Can Act As Organic Reducing Pills. The Enzyme-Catalyst Diet Program is beamed at correcting your metabolism and adjusting your "inner body clocks" so that a steady "ticking" promotes a feeling of well being, while the weight is being catalyzed out of your cells. There are many raw foods which can have the same metabolic-adjustment reaction as drugs. *The major difference is that these raw foods are natural!* Here is a selection of 25 of these foods. They can act as organic reducing pills, through their enzymatic powers.

#1–Almonds

Raw enzymes in almonds take up its calcium and help regulate mineral balance to control the appetite. Remove the skin before eating as it may be too tart to taste. Chew vigorously on several almonds to create an enzymatic flow that will offer satisfaction and appetite regulation.

#2–Apples

These enzyme fruits were used by Betty H. who was unable to take drugs and needed some natural source to control her appetite. She found that she could enjoy half portions of everyday foods, if she would eat apples whenever the hunger urge seized hold of her. The secret here is that enzymes in fresh raw apples used its phosphoric acid and minerals to stabilize the blood sugar levels so that the gnawing urge to eat could be controlled. Betty H. lost over 40 pounds while eating apples as "organic reducing pills" and continuing on with her favorite foods . . . in smaller portions. Apple enzymes made her youthfully slim again.

#3—Brazil Nuts

A prime source of nut protein which is taken up by enzymes and metabolized at a very slow rate. This creates a reduced desire for eating. Helps to regulate the body's own "self-timing" devices that signal a desire to eat.

#4—Bananas

Enzymes in the banana take its pectin and fiber network to create healthful gastrointestinal bulk and a feeling of fullness. Eat a banana as a natural appetite-tamer.

#5—Chestnuts

Gary F. is always overeating. As a busy salesman who is travelling on the road, he cannot follow any closely supervised diet. Gary F. also loves to eat. How can he bring down his heavy paunch? He does this by taking along a bag or two of roasted chestnuts. The roasting does not disturb enzyme content since the thick shell seals in the valuable elements. Whenever Gary F. has the urge to eat, either between meals, or too much food during meals, he prepares for it by eating several chestnuts. The enzymes in the chestnut take up its fibers and create *natural cellulose* in the digestive system. The chewing of the chestnuts offers oral satisfaction. The feeling of cellulose in the system offers digestive satisfaction. The urge to overeat is naturally controlled. This program took *six inches* off Gary F.'s waistline. More will be shed as he uses chestnuts with their enzyme cellulose power to control his appetite.

#6—Berries

Seasonal berries are prime sources of both citric and malic acids. These are taken up by enzymes in the berries and digestive system and used to metabolize accumulated fats and calories. Enzymes also use these natural acids to regulate blood sugar so that the appetite urge is naturally controlled. Healthy and tasty as natural organic reducing pills. Eat berries by themselves or with some yogurt as a natural appetite controller.

#7—Dates

A prime source of sun-ripened enzymes which use its natural dextrose to create a feeling of appetite satisfaction and digestive fullness. Enzymes will also use its protein to create a slow, steady burning that eases the urge to eat.

#8—Figs

An excellent natural orgainc reducing aid. Its enzymes will take the seeds and create a "bulk feeling" in the digestive system so that an excessive eating urge is reduced and there is less of a desire to want to snack.

#9—Hickory Nuts

A high protein food, it is also a good source of enzymes which can metabolize nutrients and promote a feeling of fullness and satisfaction. Its enzymes will use protein for metabolism and appetite satisfaction.

#10—Pignolia or Pine Nuts

A prime source of protein, but also contains simple sugars which are used by enzymes to promote better carbohydrate metabolism. This regulates the appetite and helps give a feeling of fullness, too.

#11—Grapes

Its fructose supply is a valuable source of energy for the body. Enzymes use grape fructose to give you "pep" and "vigor" so there is no need to want to backslide and use food as a substitute for weakness. This metabolic process also helps burn up calories and carbohydrates while putting a "stopgap" on the appetite.

#12—Nectarines

Enzymes will use the supply of bioflavonoids and vitamins from this citrus fruit to help metabolize accumulated weight in the cells. They will also alert the circulatory-metabolic systems to promote a feeling of satisfaction and appetite control.

#13—Pistachio Nuts

The greener they are, the higher in enzyme content they will be. Its enzymes take up other ingredients to form an alkaline reaction in the system to create a soothing effect that reduces the "nervous hunger" urge. Pistachio nuts also contain cellulose which is gathered up by enzymes and increased in size to form a bulk feeling. Eat a handful of pistachio nuts and you may not only want to cut down on portions of food, but eliminate extra courses, too.

#14—Walnuts

Enzymes take up the protein as well as the cellulose to form a feeling of satisfaction. Chew very carefully to extract much of the enzyme potency.

#15—Peanuts

Not really a nut, but botanically speaking, a member of the legume (bean) family. Because the peanut grows underground, it is ranked very high in biological value as a good enzyme food. Minerals from the soil "bathe" the peanut and later become part of this food. Enzymes will use these minerals for better carbohydrate metabolism and regulation of blood sugar which is the natural way to control the appetite.

#16—Lettuce

This food contains properties that help control the activity of the adrenal cortexin its responsibility of secreting the adrenalin hormone to maintain body balance. Enzymes in lettuce create this glandular balance that also helps control appetite.

#17—Onions

Raw onions contain enzymes as well as iodine which will soothe the thyroid gland, often involved in appetite control. A chopped raw onion and lettuce salad, together with a scoop of cottage cheese or buttermilk, will be a healthful "thyroid food" that can regulate the secretion of thyroxin and control the eating urge.

#18—Garlic

Edna MacM. found it impossible to control her appetite. She was more than 37 pounds overweight. It showed on her hips and her stomach. Her thighs swelled up, too. Diet drugs made her seriously ill. She needed a natural "reducing pill." She discovered garlic—but used it with parsley to neutralize its usually anti-social odor. Edna MacM. would dice a few cloves of garlic, mix with raw green sliced peppers, add a tomato, sprinkle with wheat germ, some apple cider vinegar and oil, and use this for a salad. Results? She found her appetite subsiding. She gave up her urge to nibble and snack all the time. Soon, she lost her 37 unsightly overweight pounds. She slimmed down her torso so that she now looked and felt much better. The secret? Garlic enzymes contain ethers (volatile substances) that are naturally potent and help dissolve accumulations of fats, carbohydrates and calories in the adipose cell tissues. When they melt away wastes, they also create intestinal balance, eliminating parasites so that there is less of a desire to eat. Once intestinal metabolism is regulated, appetite is better controlled. Edna MacM. uses garlic (with parsley) whenever she feels she has a compulsive urge to eat. Gradually, she is winning the battle of the bulge.

#19—Sunflower Seeds

A self-contained food which has almost all known nutrients. Enzymes in sunflower seeds take up its unsaturated fatty acids and use these to coat and lubricate the digestive organs so that you feel "filled up" but not "fattened out." Snack on sunflower or "polly" seeds whenever you want to over-eat.

#20—Whole Grains

Wheat germ is a good source of enzymes which will use its supply of unsaturated fatty acids to create a satiety value in your digestive system. Sprinkle wheat germ atop a raw vegetable salad and it becomes more than a meal in itself. It becomes an appetite-suppressor and helps ease the eating urge.

#21—Fermented Foods

Either yogurt, buttermilk, kefir milk (cultures are sold at most health stores) or cheeses made from these products are brimming with weight-melting enzymes. They use the natural fats in the products to create a coating action on the digestive apparatus so that the feeling of contentment erases the urge to eat and gain weight.

#22—Celery

Enzymes in celery will take out its natural sodium and use it to break down stubborn clumps of fat in the adipose cell tissues. Chewing crisp celery alerts mouth and digestive enzymes to correct metabolism and control eating urges. Celery also is a strong fibrous food which creates a natural bulk feeling that also eases the desire to overeat.

#23—Cucumber

A natural diuretic. Its enzymes promote the secretion and flow of urine so that there is less retention of uric acid in the system. This helps bring down edema-caused water gain. Chewing sliced cucumbers will help ease the eating urge while promoting better weight control.

#24—Carrot

Contains fiber and roughage which is used by enzymes to form a comfortable "fullness" in your digestive system and ease as well as erase the urge to want to eat heavily. Chew carrots very thoroughly. This alerts salivary glands to the digestive tasks ahead. One or two well chewed carrots can often be so "filling" that it can cut eating in half portions and create a natural reducing benefit.

#25—Apple Cider Vinegar-Oil-Honey Combo

Mix one tablespoon of pure apple cider vinegar in a glass of any desired fruit or vegetable juice. Stir in two tablespoons of polyunsaturate vegetable oil. Then add a half teaspoon of honey. Stir vigorously. You may prepare a bottle or two of this, in advance. Drink a half glass whenever you feel a compulsive eating urge. The enzymes in the apple cider vinegar take up its potassium supply, add the essential fatty acids from the oil and help to soothe and pamper your digestive system so there is less of a nervous eating compulsion. The enzymes also use minerals in the honey to satisfy your blood sugar. Natural sugars in the honey are metabolized by the enzymes to balance your blood sugar levels to protect against runaway appetites.

How To Use Organic Reducing Pills. Try all of these 25 natural enzyme foods, one a day, throughout an average four week span. Note which of these "organic reducing pills" ease your eating urges and act as natural dietetics. Then use these selected enzyme foods as part of your Enzyme-Catalyst Diet Program foods to put a natural control on your appetite and also to help you slim down permanently!

In A Nutshell:

1. Chemicalized reducing pills can be dangerous to your health so should be avoided.

2. Raw foods can act as organic reducing pills by controlling your appetite and giving you food satisfaction without adding weight.

3. Betty H. lost over 40 pounds while eating apples as "organic reducing pills."

4. Salesman Gary F. lost weight (6 inches off his waistline) on a chestnut program.

5. Edna MacM. used garlic to control her appetite and lose over 37 pounds and slim down her hips, stomach and thighs.

6. You have a tasty choice of 25 everyday foods, found in your local market, that can put a natural stopgap on your compulsive eating and create enzymatic melting of fat, carbohydrates and calories. It's a delicious way to take weight off, permanently!

11

How Proteo-Zymes
Protect Against
Flabby Skin
While Slimming Down

Simple, everyday protein foods, taken in combination with enzyme foods, can help keep your skin firm and smooth while you slim down. The Enzyme-Catalyst Diet Program calls for rebuilding your billions of body cells with a substance that "firms up" and "plumps up" your skin, to protect against the flabbiness that may result from ordinary diet plans. This substance is *collagen*. It is a "cement-like" component of connective tissue that is formed when you eat a *combination* of protein and enzymes in healthy, everyday foods. The collagen is used by your metabolic system to bolster your skin as fat is sloughed out of your cells under the ECD Program. Proteo-zymes rush in to keep your fat-reduced cells firm and youthful, so that unsightly wrinkles, "crepe" skin, aging folds, furrows can be avoided. It's the natural way to look youthful while slimming down.

THE ANTI-WRINKLE POWER OF PROTEO-ZYMES

Protein + Enzymes = Youthful Skin. Conventional diets not only made Annette V. feel hungry and weak, but they gave her skin an aging or wrinkled look. As she lost weight, she lost her firm skin and crow's feet appeared under her eyes, around her cheek bones, around her mouth, too. Annette V. developed so-called flabby skin. She lamented that along with the weight loss, she had "skin loss" that made her look much older than she really was. It was so distressing, Annette V. gained back unhealthy pounds. She looked fat, but her skin was firm. It was a vicious circle until she tried the ECD Program. This time, as she lost weight, she maintained a firm and youthful skin that was smooth and wrinkle-free. Annette V. followed this simple 3-step proteo-zyme plan:

1. *Proteo-zyme Tonic.* Three times a day, drink a glass of any fruit juice in which you stir one or two heaping tablespoons of unflavored gelatin. **Benefit:** Fruit enzymes will take the protein from the gelatin and use it to create collagen which will firm up and plump up the network of cells beneath the skin and guard against flabbiness while fat is being sloughed out of the body.

2. *Proteo-zyme Skin Cell Food.* With any desired protein food, eat a plate of fresh raw vegetables. It is important to eat the protein with the raw vegetable at the same time. **Benefit:** Vegetable enzymes will take the vigorous protein and metabolize it into amino acids which are then used to stimulate the DNA-RNA cellular substances that give birth to healthful and youthful skin cells.

3. *Proteo-zyme Nightcap.* An hour before going to sleep, drink a glass of tomato juice in which you have stirred one heaping tablespoon of Brewer's yeast. This is a high protein plant food, available at most health stores and larger supermarkets. **Benefit:** Tomato enzymes will activate the protein in Brewer's yeast to invigorate your metabolism so that as the fat is drained out of your cells, the collagen manufactured by this proteo-zyme combination will regenerate them and keep them firm. This helps keep your skin firm. **Bonus-Benefit:** The *Proteo-zyme Nightcap* was a favorite used by Annette V. because it worked overnight. *While she slept,* the *Proteo-zyme Nightcap* worked to replenish the drained out cells, using the cement-like collagen to protect against tissue breakdown and collapse. When she awoke in the morning, her skin was firm and smooth.

This easy-to-follow 3-step program using a combination of protein and enzymes on the ECD Program helps to keep the skin firm and clear and anti-wrinkling while you lose weight.

HOW TO PLAN YOUR PROTEO-ZYME SKIN SAVING PROGRAM

Daily, you should have a supply of adequate protein of approximately 100 grams. Plan to eat an abundance of fresh, raw fruits and vegetables each day. In combination, the proteo-zymes will nourish the cell membrane or "wall" in which your skin is given its youthful and anti-wrinkle appearance. The proteo-zymes will also stimulate the function of the DNA-RNA factors. These are molecules that are used to create new cells to replace those that are broken down during the weight losing program. You need a *combination* of protein and enzymes at the same meal for this process to be alerted.

High-Protein, Low-Calorie Program. Nourish your body with high-protein and low-calorie foods. The following chart shows popular, tasty foods with their protein and calorie counts. To plan your proteo-zyme program, select a balance of good protein and low calorie foods, eaten with fresh raw fruits and vegetables (and juices) daily. Your skin will be nourished as the fat is drained out of your cells. You will help protect against flabby skin while slimming down on the proteo-zyme program.

YOUR PROTEIN AND CALORIE FOOD COUNTER

FOOD	Amount (cooked weight)	Protein (grams)	Calories
MEAT†			
BEEF			
Pot Roasts:			
Arm	4 oz.	40	316
Blade	4 oz.	42	358
Heel of round	4 oz.	37	265
Rolled neck roast	4 oz.	37	295
Roasts:			
Sirloin tip	4 oz.	35	223
Standing rib	4 oz.	27	342
Standing rump	4 oz.	38	282
Steaks:			
Bottom round steak	4 oz.	43	286
Club steak	4 oz.	31	336
Flank steak	4 oz.	40	282
Porterhouse steak	4 oz.	30	290
Rib steak	4 oz.	31	314
Sirloin steak	4 oz.	31	250
T-bone steak	4 oz.	30	296
Tenderloin	4 oz.	31	269
Others:			
Brisket	4 oz.	31	370
Ground beef, chuck	4 oz.	31	316
round	4 oz.	36	196
Short ribs	4 oz.	29	484
Stew, chuck	4 oz.	27	505
round	4 oz.	39	312
LAMB			
Chops:			
Arm	4 oz.	31	302
Blade	4 oz.	33	336
Loin	4 oz.	33	268
Rib	4 oz.	31	349
Riblets	4 oz.	28	479
Roast leg	4 oz.	34	234
PORK, CURED			
Roast ham, butt	4 oz.	30	246
Roast ham, shank	4 oz.	31	281
Smoked boneless shoulder, butt	4 oz.	27	382
PORK, FRESH			
Chops and steaks:			
Blade steak	4 oz.	35	332
Leg, center slice	4 oz.	44	284

Roast:			
Boston butt	4 oz.	30	340
Fresh picnic shoulder	4 oz.	31	295
Sirloin roast	4 oz.	36	272
Tenderloin	4 oz.	37	287

PORK, PROCESSED

Bacon, Canadian-style	1 slice	7	62
Bacon, regular	1 strip	2	34
Bacon, thick slice	1 strip	3	54
Sausage link	1	3	63

VEAL

Loin chop	4 oz.	41	248
Roasts:			
Standing rump	4 oz.	37	200
Sirloin	4 oz.	34	211
Steaks:			
Arm	4 oz.	43	240
Blade	4 oz.	40	253
Cutlet	4 oz.	46	242
Rib	4 oz.	40	258
Sirloin	4 oz.	42	245
Stew	4 oz.	34	415

VARIETY MEATS

Heart	4 oz.	35	235
Liver, beef	4 oz.	32	221
lamb	4 oz.	41	268

pork	4 oz.	36	244
veal	4 oz.	37	272
Tongue	4 oz.	22	340

SAUSAGES AND COOKED SPECIALTIES

Bologna (1 slice)	1 oz.	4	66
Frankfurter (1)	1 2/3 oz.	7	124
Liver sausage (1 slice)	1 oz.	5	79
Luncheon meat (1 slice)	1 oz.	5	81
Salami (1 slice)	1 oz.	7	130
Vienna sausage	1 oz.	5	65

FISH

Clams, canned	3 oz.	7	44
Cod, dried	1 oz.	23	106
Crab, canned	3 oz.	14	89
Haddock, fried	3½ oz.	19	158
Halibut, broiled	4½ oz.	33	228
Lobster, canned	3 oz.	16	78
Mackerel, canned	3 oz.	17	154
Oysters, raw	½ cup	12	100
Salmon, pink, canned	3 oz.	17	122
Salmon, broiled	4 oz.	34	204
Sardines, solids only	3 oz.	19	177
Shrimp, canned	3 oz.	23	108
Swordfish, broiled	4½ oz.	34	223
Tuna, canned	3 oz.	25	169

POULTRY

Chicken, roasted:			
white meat	4 oz.	38	166
dark meat	4 oz.	31	202
Turkey, roasted:			
white meat	4 oz.	41	246
dark meat	4 oz.	37	273

EGGS

Egg	1 med.	6	77
Egg, scrambled or fried	1 med.	7	106

DRIED BEANS AND PEAS; NUTS

Beans, canned or cooked red kidney	½ cup	7	115
with pork and tomato sauce	½ cup	8	148
Beans, lima, dry (uncooked)	½ cup	19	305
Coconut, dried, shredded	¼ cup	1	86
Peanuts, roasted, shelled	¼ cup	10	201
Peanut butter	1 tbsp.	4	92
Peas, split, dry (uncooked)	½ cup	25	345
Pecans, shelled	¼ cup	3	188
Walnuts, shelled	¼ cup	4	166

FRUIT

Apple	1 med.	Tr.	76
Applesauce, sweetened	½ cup	Tr.	92
Apricots,			
fresh	3	1	54
canned, water pack	½ cup	1	39
canned, syrup pack	½ cup	1	103
Banana	1 med.	1	88
Berries, fresh (black-berries, blueberries, raspberries)	½ cup	1	42
Cantaloup	½	1	37
Cherries,			
fresh	½ cup	1	33
canned, sour	½ cup	1	61
Cranberry sauce	¼ cup	Tr.	137
Fruit cocktail, canned	½ cup	1	90
Grapefruit,			
fresh	½ med.	1	75
sections	½ cup	1	39
Grapefruit juice, canned, unsweetened	½ cup	1	46
Grapes, fresh	½ cup	1	47
Orange	1 med.	1	70
Orange juice, fresh or frozen	½ cup	1	54

Food	Amount		
Peaches,			
fresh	1 med.	1	46
canned, water pack	½ cup	1	33
canned, syrup pack	½ cup	1	87
Pears,			
fresh	1 med.	1	95
canned, water pack	½ cup	Tr.	38
canned, syrup pack	½ cup	Tr.	87
Pineapple,			
fresh	½ cup	Tr.	37
canned, syrup pack	Sm. sl. with syrup	Tr.	48
Pineapple juice, canned	½ cup	Tr.	61
Plum, fresh	1	Tr.	29
Prunes, cooked, unsweetened	½ cup	1	155
Prune juice, canned	½ cup	1	85
Strawberries, fresh	½ cup	1	27
Watermelon	½ sl. (¾" x 10")	1	45

VEGETABLES

Food	Amount		
Asparagus	½ cup	2	20
Beans, lima	½ cup	4	76
Beans, green or wax	½ cup	1	14
Beets	½ cup	1	34
Broccoli	½ cup	3	22
Brussels sprouts	½ cup	3	30
Cabbage, raw	½ cup	1	12
cooked	½ cup	1	20
Carrots, raw	½	Tr.	11
cooked	½ cup	Tr.	22
Cauliflower	½ cup	1	15
Celery, raw	½ cup	1	9
cooked	½ cup	1	12
Collards	½ cup	4	38
Corn (5-inch ear)	1	3	84
canned	½ cup	2	70
Cucumber slices	6	Tr.	6
Green pepper	½ med.	Tr.	8
Kale	½ cup	2	23
Lettuce	1/6 head	1	11
Okra	8 pods	2	28
Onions, raw green	6 sm.	1	23
cooked	½ cup	1	40
Peas	½ cup	4	64
Potato, baked or boiled	1 med.	3	107
Radishes	4 sm.	Tr.	4
Rutabagas	½ cup	1	25
Sauerkraut	½ cup	1	16
Spinach	½ cup	3	23

BREADS AND CEREALS

BREADS

Food	Amount		
Biscuit (2½" round)	1	3	129
Cornbread (2¾" round)	1	3	106
Crackers, graham and soda	2	1	51
rye	2	2	43
Doughnut	1 med.	2	136
Muffin (2¾" round)	1	4	134
Pancakes (4" round)	3	5	177
Roll, plain	1 med.	3	118
sweet	1 med.	5	178
Rye bread	1 slice	2	57
Waffle (4½ x 5 5/8 x ½")	1	7	216
White bread	1 slice	2	63
Whole wheat bread	1 slice	2	55

BREAKFAST CEREALS

Food	Amount		
Bran, flakes (40%)	½ cup	2	58
whole	½ cup	4	43
Corn flakes	½ cup	1	48
Farina, cooked	½ cup	2	52
Oatmeal, cooked	½ cup	3	74
Rice, flakes	½ cup	1	59
puffed	½ cup	Tr.	23
shredded (1-oz. biscuit)	1	3	102

Food	Amount		
Squash, summer	½ cup	1	17
winter	½ cup	2	43
Sweet potato, baked	1 med.	3	183
Tomato, raw	1 med.	2	30
canned or cooked	½ cup	1	23
Tomato juice	½ cup	1	25
Turnips	½ cup	1	21

DAIRY FOODS

MILK

Food	Amount		
Buttermilk	1 cup	9	86
Chocolate-flavored milk drink	1 cup	8	185
Cocoa, made with milk	1 cup	10	236
Cream, light	1 tbsp.	Tr.	30
whipping	1 tbsp.	Tr.	49
Dry milk solids, non-fat	1 tbsp.	3	28
Evaporated milk	½ cup	9	173
Skim milk	1 cup	9	87
Whole milk	1 cup	9	166

CHEESE

Food	Amount		
Cheddar or Swiss	1 oz.	7	109
Cottage	¼ cup	11	54
Cream cheese	1 tbsp.	1	56

OTHER CEREAL PRODUCTS

Macaroni, spaghetti, cooked	½ cup	4	107
Noodles, cooked	½ cup	2	54
Rice, white, cooked	½ cup	2	101

DESSERTS

Cakes, angel food or sponge (8")	2" wdg.	3	113
rich, with plain icing (6")	3" wdg.	4	378
iced cupcake (2¾''')	1	3	161
Cookie, plain (3")	1	2	109
Custard, baked	½ cup	7	142
Gelatin, plain	½ cup	2	78
with fruit	½ cup	2	86
Ice cream, plain	¼ pint	3	147
Pies, apple (9")	4" wdg.	3	331
custard or pumpkin (9")	4" wdg.	6	265
lemon meringue (9")	4" wdg.	4	302
Pudding, vanilla	½ cup	4	138
Sherbet	¼ cup	1	119

SOUP

Bouillon or consomme	1 cup	2	9
Chicken	1 cup	4	75

Cream soups (asparagus, celery or mushroom)	1 cup	7	201
Tomato	1 cup	2	90
Vegetable	1 cup	4	82

PICKLES, OLIVES, CONDIMENTS

Catsup or chili sauce	1 tbsp.	Tr.	17
Olives, "mammoth" size, green	5	Tr.	36
ripe	5	1	53
Pickles, sweet (2")	1	Tr.	11
dill (4")	1	1	15

FATS, OILS, SALAD DRESSINGS

Butter or margarine	1 pat, 1½ tsp.	Tr.	50
Fresh salad dressing	1 tbsp.	Tr.	59
Mayonnaise	1 tbsp.	Tr.	92
Oils, salad or cooking	1 tbsp.	0	124

SUGAR, CANDY, OTHER SWEETS

Candy, hard, fudge or caramels	1 oz.	Tr.	114
milk chocolate	1 oz.	2	143
peanut brittle	1 oz.	2	125
Honey	1 tbsp.	Tr.	62
Jams, jellies, preserves	1 tbsp.	Tr.	53

153

Molasses, light	1 tbsp.	0	50
Syrup, table blends	1 tbsp.	0	57
Sugar, granulated	1 tsp.	0	16
brown	1 tsp.	0	17

BEVERAGES

Alcoholic,			
beer	12 oz.	1	171
gin, rum, whiskey	1 oz.	0	77
wine	2 oz.	0	65
Carbonated,			
cola-type	1 cup	0	107
ginger ale	1 cup	0	80
Coffee, black,			
unsweetened		0	0
Tea, plain		0	0

FOUNTAIN SPECIALS*

Banana Split	1 avg.	11	600
Chocolate Ice Cream			
Soda	1 avg.	5	390
Chocolate Sundae	1 avg.	5	362
Chocolate Malted Milk	12 oz.	11	419
Chocolate Milk Shake	12 oz.	10	398

*Based on standard soda fountain formulas distributed by the International Association of Ice Cream Manufactures.

†The figures for meat are based on research conducted at the Oklahoma Agricultural Experiment Station.

HOW TO BUDGET YOUR "CALORIE EXPENSE ACCOUNT"

After you know the approximate number of calories you will need each day to allow you reasonable, yet comfortable weight loss, (See Chapter 6 for various calorie charts), keep a daily "calorie expense account" budget. This tells you how you are progressing.

Freda R. Loses Weight On A Scheduled ECD Program. Before Freda R. began dieting, she weighed 180 pounds. She needed the self-discipline of a regular scheduled weight loss program. She set herself a goal of losing 60 excess pounds over a four week period. She wanted to become a slim 120 pounds. When she started the ECD Program, she prepared a chart to show how many calories she ate daily and how much she lost at the end of each week. She would weigh herself regularly. She felt more motivated as she saw how the ECD Program began to melt away pounds and she stuck to the plan. At the end of the four week schedule, she had lost the 60 excess pounds and was now a slim 120. This form of self-discipline and self-motivation is easy with a chart. Here is Freda R.'s chart and how she kept writing down her progress:

FREDA'S RECORD OF CALORIES IN FOOD EATEN
and
WEIGHT LOSS EACH WEEK

Date _____ Present Weight _____ Desired Weight _____

Day	Breakfast	Lunch	Dinner	Extras	Total Calories for Day	
1						
2						Weight at end of
3						week _____
4						
5						Total weight loss to
6						date _____
7						

8					
9					
10					
11					
12					
13					
14					
15					
16					
17					
18					
19					
20					
21					
22					
23					
24					
25					
26					
27					
28					

Weight at
end of
week _____

Total weight
loss to
date _____

Weight at
end of
week _____

Total weight
loss to
date _____

Weight at
end of
week _____

Total weight
loss to
date _____

Here's how you can make up your own chart of calories in foods you eat and your weight loss. Just copy Freda R.'s chart on a sheet of paper without her entries. Use this chart to record your own weight-losing progress on ECD Program.

Egg + Orange Wedges = Natural Face Lift. Harriet DeB. had tried one diet after another. While many left her weak, she did lose weight . . . but her youthful skin was "lost," too. She was "starving" her body and "starving" her skin. With deep wrinkles, a sagging throat, hollow cheeks, her skin looked as if it were suffering from malnutrition. Indeed, it was. Harriet DeB. was faced with a choice of either looking old and slim, or having a youthful skin while being overweight. She found a solution to this dilemma with the ECD Program. But she added something else. Three times a week, for breakfast, she would have a soft boiled egg and toast . . . with a huge bowl of orange wedges. In combination, she had a proteo-zyme reaction wherein the protein of the egg was used by the enzymes from the fruit to nourish the two principal layers of her skin. That is, the *dermis* which is composed of "uplifting" elastic connective tissues; also, the proteo-zymes nourished the *epidermis* so that broken, damaged or decaying cells could be quickly replaced and her skin could be protected from sagging.

Benefits of Egg-Orange Combination: The egg is considered a nearly perfect food. It was reported,[1] "Because of the percentage and quality of protein present, eggs are classed as one of the most important protein foods in the diet. Egg protein is a *complete protein;* it contains all of the essential amino acids which are required by the body to build and renew body tissues. In fact, egg protein is of such high quality that it is the standard against which the quality of other food protein is measured."

The orange is a prime source of fructose and levulose, both of which are natural energizing enzymes. These substances activate and amplify the vigor of the *complete protein* in the egg so that it can nourish and rebuild the *complete* components of the skin cell to guard against tissue breakdown often caused by loss of supportive elements during weight reduction.

This *combination* acted as a "natural face lift" for Harriet DeB. She lost weight, soon shed some 58 unwanted pounds on the Enzyme-Catalyst Diet Program. But just as important this Egg-Orange Combination kept her skin cells proteo-enzyme nourished so

---

[1] *Parker Natural Health Bulletin*, West Nyack, New York 10994. Vol. 4, No. 26, December 23, 1974. Available by subscription.

that she looked youthful with no risk of deep wrinkles, sagging throat or hollow cheeks.

Secret Of Proteo-Zyme Combination: The *complete* protein offers "total nourishment" to the skin cells so that it can help protect against wrinkling and tissue breakdown. The enzymes can function to their fullest capacity in tissue nourishment when given *complete* protein as found in the egg. Just three eggs weekly with enzyme-high orange juice can offer the skin tissues this needed nourishment and protection against premature aging while slimming down.

25 PROTEO-ZYME SKIN TONICS FOR LASTING YOUTH AND HEALTH

Here is a variety of proteo-zyme skin tonics that invigorate the metabolism so that protein is used by enzyme catalysts to nourish the skin and keep it looking youthful and healthy while you slim down.

1. Spread a mixture of raw egg white and lemon juice over your face and neck—not your eye area. Let remain until very dry. Rinse off in lukewarm and then cold water. **Benefit:** Helps firm up skin tone, close up pores, alert a sluggish circulation to give you a fresh, clean look.

2. Slice or grind a cucumber. Mix with very heavy dairy sour cream in a blender. Apply to the skin. Rub vigorously in circular motions. Rinse off in lukewarm and cold water. **Benefit:** Helps cleanse clogged pores and restore moisture to the depleted skin cells through enzyme-protein action.

3. Deep-clean your pores by simmering herbs in water. Then lean over this fragrant brew with a tent-like towel draped over your head and the steaming pan. **Benefit:** Enzymes in the herbs use body protein to create rebuilding of the cells and tissues.

4. Treat your skin to an enzymatic milk wash once a day for better color. Use either milk, yogurt or even buttermilk. Add some fruit juice. Then coat the face with this mixture. Massage gently. Rub deep into your skin. Then wash off in contrasting warm and cool water. **Benefit:** Enzymes use the milk protein to replace moisture loss beneath the skin surface to help protect against dryness which is the forerunner of wrinkling. Boosts skin color, too.

5. Sallow skin is often caused by weak protein metabolism during dieting. To guard against this skin aging reaction, try a cucumber cleansing. Rub a sliced cucumber all over your face. Or, mash a cucumber and gently massage it into your face. **Benefit:** Enzymes in the raw cucumber activate the protein in your skin cells to boost a sluggish metabolism. This puts color into otherwise sallow skin, during slimming down.

6. Combine strawberries with whole milk until you have a slight mush. Rub all over the face and arms or anywhere that you are troubled with wrinkled skin. Let remain up to 60 minutes. Then sponge off with alternating warm and cool water. **Benefit:** Enzymes in the berries will take the casein protein from milk and use it to help restore a natural acid-alkaline balance to protect against premature aging of the skin.

7. Mix one half avocado with two tablespoons of honey. Now blend with enough whole milk or cream until of a cold cream mixture. Use nightly as a cleansing cream. **Benefit:** The avocado enzymes take up the unsaturated fatty acids within this fruit and then blend it with the levulose of the honey. This combination now activates the protein in the milk to nourish the deep reservoirs in the skin pores, so they are firm and healthy and wrinkle-resistant.

8. Rub slices of raw potato against any parts of your skin that are oily. Let dry overnight. Next morning, wash off. **Benefit:** While fat is lost from your body on the ECD Program, the skin may become oily. Fat cells shed the oil through the pores. The raw potato is a good source of enzymes which absorb this oil; at the same time, potato enzymes enter through the skin pores and activate the body's supply of protein to guard against excessively oily skin during the weight loss program.

9. Blend two egg yolks (complete protein) with two tablespoons of liquid glycerine (sold at most pharmacies) and apply to your face as a 30-minute facial. Then rinse off in contrasting warm and cool water. **Benefit:** Enzymes in the raw egg yolks will take its supply of complete protein, then "slide" through the pores on the glycerine, where a moisturizing of the skin cells helps "plump up" your thirsty reservoirs of skin cells and guard against furrows and crow's feet.

10. If soap irritates your skin, then make your own proteo-zyme "soap" which cleanses and nourishes at the same time. Mix oatmeal and honey into a gooey paste. Rub this all over your face, just as you would creamy soap. Then wash off with a soft cloth. You may also want to put sparkle into your skin by finishing with a sparkling seltzer rinse. That's right. Rinse off with seltzer water. It tones up your skin and makes it look fresh and alive. **Benefit:** Enzymes in the honey take up the grain protein from the oatmeal and use it to

replenish lost moisture from beneath the skin cells. Enzymes use the polyunsaturated fatty acids with the protein in oatmeal to create this natural skin feeding benefit. The seltzer alerts sluggish pools of toxic debris "locked" in pockets beneath the skin surface. This helps distribute the toxic debris and exhilarate the enzymatic reaction for better skin health.

11. Beat two egg yolks in a bowl. Slowly, add a half cup of polyunsaturated vegetable oil. When this mixture starts to solidify, then add several tablespoons of apple cider vinegar. Mix thoroughly. Use it to remove dirt from your skin as a soap substitute. Finish with a sparkling cool water rinse. **Benefit:** The vinegar enzymes carry the egg protein on a "sea" of fatty acids from the oil, into the pores of the skin. Here, this proteo-enzymatic action is to cleanse away clogged impurities and dissolve them. Once the debris is cleansed, then the skin can become firm and smooth and resist wrinkles.

12. Put grape juice over your face, especially in those areas that show sagging or a tendency toward wrinkling. After it dries, rinse off with cool water. **Benefit:** Grape enzymes are especially invigorated by this fruit's fructose-levulose energizers. The enzymes seep into the skin and help dissolve accumulated wastes that are often left over after fat combustion during weight loss. By washing away these wastes, the skin becomes smoother and less furrowed, as pounds are melted away.

13. Treat yourself to a body proteo-enzyme soak that will make your skin glow with youthful health. Simmer one cup of ordinary barley in 5 cups of water in a covered saucepan for 30 minutes. Strain. Now add lemon and orange rinds. Let cool and soak for 60 minutes. Now add the juice of several lemons and grapefruits. Mix together. Pour into a tub of lukewarm water. Soak yourself luxuriously for 60 minutes. Then let the water drain out. Finish with a cool fresh shower spray. **Benefit:** The fruit enzymes will take up the barley protein, and in the pore-opening warmth of the bath, send a wave of internal nourishment to your billions of body cells. This helps keep them repaired and youthful so they will not collapse and cause body wrinkling or aging from fat loss.

14. Combine several tablespoons of yogurt with cream. Apply wherever you see wrinkles or crow's feet. Wait 30 minutes until it is dry. Then rinse off with splashes of cool water. **Benefit:** Enzymes in the fermented milk transport fatty acids from the cream throughout your epidermis to moisturize the cells and cause DNA-RNA regeneration. This maintains better tissue integrity to protect against skin aging during reducing.

15. Make a paste or a mash of slightly steamed carrots, turnips, radishes, beets. When smooth and consistent, spread over skin portions that show creases or blemishes. Let remain up to 45 minutes. Rinse off with contrasting warm and cold water. **Benefit:** Your body protein will be activated by the energizing effects of the carbo-zymes to create a synergistic reaction wherein your dermis self-regenerates its cells to guard against wrinkling and so-called "reducer's aging."

16. Beat the white of an egg. Into this, fold 4 tablespoons of freshly squeezed grapefruit juice. Apply wherever you notice crow's feet, especially around your mouth, chin, throat hollow. Let remain for 30 minutes. Wash off with contrasting warm and cold water. **Benefit:** Strong astringent-like grapefruit enzymes whip up the pure egg white protein and "zing" it within your skin pores. This proteo-zyme reaction boosts sluggish metabolism so that your skin cells are nourished; collagen is built to give firmness and plumpness to your skin.

17. A housewife, Jacqueline Z., was more than 40 pounds overweight when she put herself on a "starvation diet." Her extra pounds dropped away, but at the same time, her formerly "stretched" skin now sagged and she developed a "turkey gizzard" in her throat hollow that made her look like an old woman. When Jacqueline switched to an Enzyme-Catalyst diet, the enzymes washed the excess fat out of her cells, which slimmed her down . . . and, more important, *kept her slimmed down.* To protect against aging skin, she followed a simple proteo-zyme program: after preparing scrambled eggs, she put aside *the white shell linings or membranes of the eggs.* While these linings were still moist, Jacqueline Z. applied them with her fingertips all over visible lines. This included her forehead, nose-mouth indentations, furrows around the eyes. She let this remain for 15 minutes. Then she splashed off with warm water. Presto! Her skin remained youthfully smooth. Her unsightly sags tightened up. She glowed with firm youthfulness as she slimmed down on the ECD program. A simple program, but remarkably effective. **Benefit:** The enzymes in the eggshell lining consist of pure protein with all known amino acids. The enzymes send these amino acids into the skin cell nucleus, mobilize the DNA-RNA molecules to regenerate collagen and cause internal buildup so the skin is supported on strong structures. It is a natural anti-wrinkling program.

18. Scalp dryness can be corrected with a simple proteo-zyme remedy. Combine one ripe avocado with a cup of whole milk. Then beat in a raw egg. Combine thoroughly. Now massage this mixture deeply into your scalp. Massage vigorously. Rub your scalp around to promote better circulation. Then rinse thoroughly under contrasting warm and cool water. **Benefit:** Proteo-zymes in the avocado and milk take the fatty acids from the egg and send a stream of nourishment through your hair follicles. Replenished scalp tissues are now moisturized and dryness is avoided.

19. Draw a bath of comfortably warm water. Add two cups of apple cider vinegar. Then soak yourself for 30 to 45 minutes. Finish with a cool fresh water splash. Dry off. **Benefit:** Enzymes in the vinegar seep through skin pores and stimulate your own protein stores to improve metabolism so that the skin cells are better nourished. This helps build up epidermal moisture and protection against body wrinkling caused by dryness during loss of fat in reducing.

20. Mash cooked tapioca with some vegetable oil. Add fresh fruit juice. Mix together. Apply wherever you see trouble spots such as skin chapping, blemishes, furrows, wrinkles. Let remain for 60 minutes. Then rinse off. **Benefit:** The tapioca protein (grain source) is a prime source of fatty acids which are moisturized further by the oil and then sent "shooting" into the body by the enzymatic action of the fruit juice. Helps ease tendency toward skin aging, when traced to moisture loss because of weight reduction.

21. Apply a mixture of honey and fruit juice around the hollows of your eyes, on your eyelids, too. If possible, let remain overnight. Rinse off with cool water. **Benefit:** Fruit enzymes take the protein from the honey and "coat" the skin so that the pores "drink" the proteo-zymes and nourish the cells beneath. Helps guard against aging eye areas.

22. Combine a few tablespoons of mineral oil with a few tablespoons of whole milk. Now add one cup of cornstarch. Stir together. Pour into a warm bath. Soak up to 45 minutes. Then finish with a warm and cold water shower. **Benefit:** The proteo-zymes from the milk and cornstarch activate sluggish cells beneath the skin surface to stimulate the manufacture of collagen, needed for better health of the capillaries. Protects against skin wrinkling because of dieting.

23. Mash half a peeled avocado very well. When completely lumpless, blend it into a half cup of sour cream. (If you like a minty fragrance, add a few drops of peppermint oil, available from your pharmacist; do not use food coloring.) Apply to your clean face.

Relax for about 30 minutes. Remove with tepid water and a sponge or cloth, followed by a few clear rinsings with cool water. **Benefit:** Proteo-zymes in the avocado take up the fatty acids from the sour cream and transport them into the skin where the reaction is replenishment of lost moisture so skin cells do not "die" from "thirst." This helps keep the skin smooth and wrinkle-free during weight loss when excess fat is sloughed out of the cells.

24. Put a peeled *very* ripe banana and a peeled ripe avocado into a potato ricer. Press well so both fruits are blended. Then rub the mixture all over your body or on those areas where the skin needs feeding to combat dryness. For best results, let it remain for 60 minutes before rinsing or showering off. **Benefit:** Carbo-zymes in the banana are invigorated by the proteo-zymes from the avocado and send a mixture of nourishment and moisture to the cells and tissues of your body. This helps keep them healthy and "plump" while you slim away fat from your body.

25. Mash half of a peeled avocado with a half cup or ordinary table salt. Stand in a tub or shower and rub the mixture all over your body (or on trouble spots) with a great deal of friction. When you think you've had enough, shower off. **Benefit:** Proteo-zymes are strengthened by the abrasive power of the salt and dead skin will sluff off in a miraculous way. This smooths your skin and keeps it glowing, as you lose unwanted pounds.

A combination of protein plus enzymes can help keep your skin firm and youthful, as you slim down. Proteo-zymes nourish the DNA-RNA molecules beneath your skin and keep you looking youthful . . . inside and outside.

Highlights:

1. Protein + Enzymes = Youthful Skin. This is the reward enjoyed by Annette V. who combated "skin sag" on the ECD program.

2. A simple 3-step program helps keep skin smooth and wrinkle-free while weight is lost.

3. Plan your Proteo-Zyme Skin Saving Program with a high-protein, low-calorie and high-enzyme diet. The protein and calorie food counter offers at-a-glance counts of these foods.

4. Harriet DeB. used two everyday foods to give her skin a "natural face lift" while her body slimmed down.

5. Treat your skin to new youthfulness with a variety of the 25 proteo-zyme tonics for lasting youth and health.

6. Jacqueline Z. lost some 43 extra pounds on the ECD Diet Program, but kept her skin firm and free from "stretch" and "sag" with a simple proteo-zyme program using an everyday food item for pennies per treatment.

12

Your ECD Menu Guide For Fast, Permanent Weight Loss

Edward B. had a 48 inch waistline that would not go down. After each meal, he had to open his belt. Walking was strenuous. If something fell on the floor, he had to call for help to pick it up. If he bent over, to tie his shoe laces, he felt shortness of breath; he would wheeze and sputter with exertion as he finally managed to straighten up. He had tried conventional diets that did shrink his waistline, but make him so hungry, he soon started overeating again, and gained back his extra weight and his corpulent waistline. Edward B. had to do something. He needed a plan that would let him eat but help him reduce and stay reduced. This is where the Enzyme-Catalyst Diet Program came to be his sought-after hope for permanent weight loss.

Basic 5-Step Menu-Planning Guide. Whether eating at home or outside, Edward B. followed this 5-step planning guide:

1. *Before each meal, drink a glass of fresh, raw fruit or vegetable juice.* **Benefit:** Enzymes boost sluggish metabolism so that the fat-melting mechanism is alerted and there is a greater metabolic activity that will help burn up excess calories and fats from the billions of body cells and tissues.

2. *Each and every meal of the day should begin with a raw plant food and end with a raw plant food, preferably as a salad.* **Benefit:** When you eat fat-containing or protein-containing foods, the process of metabolic combustion will deposit calories and fats in the adipose tissues, causing weight buildup. To alert your digestive-metabolic systems to promote a burning up of these weight-causing elements, enzymes are needed as catalysts. Eating raw plant foods will send forth a supply of needed enzymes that will burn up the excess fats and help "slim down" the adipose tissues, thereby keeping the body permanently slim.

3. *Each and every meal must have a raw food that requires healthful chewing.* **Benefit:** The process of chewing stimulates the salivary glands which issue forth a group of important fat-calorie melting enzymes. This process also alerts the digestive and intestinal tracts to secrete enzymes that wait to "attack" foods and metabolize them, dissolve them to promote better assimilation. chewing also triggers off the body's enzymatic system to promote top-to-toe metabolism so that adipose cells are "washed" and fat-calorie buildup is restricted. It is the natural way to help you slim down and stay slimmed down.

4. *Enjoy a nightcap of a fresh mixed vegetable cocktail for "slim-while-you-sleep" metabolism.* **Benefit:** Raw vegetables are a prime source of carbo-zymes, or *energetic-enzymes* which are self-powered and self-propelled. They work while you sleep to help burn up accumulated fat-calorie buildup in the adipose tissues. Weight loss can be enjoyed without any effort while you sleep when carbo-zymes in raw vegetable juices create spontaneous combustion and slim down your cells—and slim down your body, too.

5. *Instead of a weight-building snack, reach for a weight-melting fruit.* **Benefit:** Fresh fruit creates a molecular-biological reaction which promotes the acceleration of metabolism so that the accumulating fats and calories can be burned off and weight can then be reduced or controlled.

Edward B. Loses 14 Inches From Waistline. On this easy 5-step program, Edward B. found that he had a natural control for his appetite. But it also enabled him to enjoy most of his favorite foods, provided he included these 5 steps with his daily meals and eating programs. He was delighted to keep taking in his belt and pants. Soon, he had measured a happy loss of 14 inches from his waistline. Most important, the weight was off . . . *permanently,* on the ECD Program. Edward B. now felt youthful, healthy and satisfied.

PLANNING ENZYME-CATALYST DIET MEALS
FOR YOUR HEALTH AND PLEASURE

A nutritionally sound diet should provide you with vitamins, minerals, protein, fats, carbohydrates and calories to maintain desirable health and weight. All foods furnish calories. Furthermore, each gram of fat furnishes two and one-fourth times as many calories as each gram of either protein or carbohydrate. The trick is to fortify yourself with *enzymes* that act as catalysts to split the molecules of the calories, carbohydrates and fats, transfering them through the metabolic system for assimilation and disposal. This will help control weight buildup. With enzymes from raw foods, you can enjoy good food, but your should follow some guidelines.

How To Help Enzymes Keep You Slim. Cooperate with enzymes by modifying your eating practices. Here is a 5-step guideline:

1. Let moderation be your guide—*don't overeat.*

2. If overweight, reduce and maintain normal weight by increasing your intake of raw foods, solids and liquids.

3. Choose a wide variety of foods from each of the food groups listed in this chapter. This offers you good taste, pleasure and enzymes for slimming down.

4. Eat less of the very high-fat foods—pastries, cakes, cookies, excessive amounts of fatty meats and high-fat dairy products. If you *must* eat these foods, then "sandwich" them with enzymes. That is, *begin with a raw fruit salad, then end with a raw fruit salad. This will give your digestive system a double-barrelled supply of enzymes needed to break down and dissolve the added amounts of calories, fats or carbohydrates.*

5. Limit "empty calorie" foods (those which contribute little, if any, nutritive value)—sugar, candy, soft drinks and other concentrated sweets. These also displace the metabolic activity, and weaken the enzymatic reaction so weight builds up. Instead of a sweet, reach for a fruit.

SIX BASIC PRINCIPLES OF ECD MEAL PLANNING

1. You may include a source of first-rate protein at every meal—such as lean meat, fish, poultry, cottage-type cheese, or skim milk. Limit eggs to three per week if concerned about cholesterol.

2. You may include generous servings of vegetables: preferably, raw vegetables. Cook those that must be cooked.

3. You may include whole grain bread and cereal products or potatoes at each meal.

4. You may include at least three fruits a day. Emphasize citrus fruits which have stronger enzymatic content.

5. You may include vegetable oils and margarines made from liquid vegetable oil for their polyunsaturated fatty acid content. Enzymes will use these substances to offer a feeling of satisfaction, while helping to dissolve away accumulated "hard" fats from adipose tissues.

6. Limit to special occasions, the eating of pastries, cakes, cookies, candy and soft drinks.

FISH, MEAT, EGGS

Fish. Enjoy fish at least 4 to 5 times a week, for breakfast, lunch or dinner. Enzymes need the fish fat for better cellular metabolism and fat-washing.

Poultry. Eat poultry often. It is low in hard fat and contains special unsaturated fatty acids that enzymes need for cellular metabolism.

Veal. Enjoy veal frequently because it is lean and is of a type that enzymes can metabolize and assimilate with less weight buildup.

Beef, Lamb. Enjoy three to four times a week. If you eat beef or lamb with a raw salad, assimilation is improved and there is less fat buildup.

Liver, Heart, Kidney. These are high in saturated fats which require stronger enzymatic action. Limit to occasional intake. When you do eat these foods, you should precede and complete the meal with raw fruits or raw vegetables.

Eggs. For cholesterol watching, limit to three per week. When eating eggs for breakfast, have a glass of fresh fruit juice with it for better metabolism of the saturated fats.

Avoid very fat meats such as bacon, sausage, corned beef, pastrami, luncheon meats. These require very powerful enzymatic metabolism and excessive intake will weaken your digestive functions.

Select lean cuts of all meats.

Trim off all visible fat.

Keep moderate servings. About 4 to 6 ounces before cooking will equal 4 ounces after cooking.

Make Friends With Your Enzymes. Your digestive system will enjoy "friendly enzymes" if they are not overworked, overtired, overexhausted. Therefore, make friends with your enzymes by giving them tasty, delicious foods that are as fat-free as possible for adequate metabolism. Do not abuse your enzymes with marbled, heavy foods which can cause fat buildup that these catalysts cannot control.

MILK AND MILK PRODUCTS

2 cups of skim milk daily for adults.

Cottage, pot or farmer cheese often. It is low in fat and high in protein.

Enzymes are available in raw fruits, vegetables, seeds, nuts, grains. Eat these last-named foods *together* with milk and milk products for better assimilation and lesser fat buildup. The old tradition of fruit with cheese is a "slimmer's secret." Eat this combination regularly.

Note: If you must eat butter, cream, ice cream, cream cheese, hard cheeses and other whole milk cheeses, then you must have a raw fruit salad or raw vegetable salad either *before* or *after* the meal. For better enzymatic assimilation, the salad should be had *both* before and after this heavy dairy meal.

FRUITS AND VEGETABLES

Fruits. Eat these daily. Either whole fruit or juice or both forms. This is the natural catalyst that will help burn up weight in tissues and cells.

Vegetables. Daily, eat a selection of fresh, raw vegetables. Enjoy cooked vegetables such as broccoli, turnip greens, kale, pumpkin, squash, sweet potatoes and all other seasonal plant foods that *must* be cooked.

WHOLE GRAIN BREADS OR CEREALS

Enjoy these daily. Eat fruit with these foods for better assimilation of carbohydrates and calories.

HOW TO PLAN YOUR THREE (OR MORE) DAILY MEALS

Breakfast: This should be substantial to nourish the body, after a night's sleep, also to provide it with a supply of enzymes needed to catalyze foods to be eaten during the day. Fresh fruit is important to bolster energy. Also have whole grain bread or cereal, skim milk, a protein food, a handful of nuts or seeds with fresh fruit for additional enzymatic vigor.

Luncheon: A raw fruit or vegetable salad. A hot dish may include soup or stew. Try barley, brown rice, lentils, lima beans, split peas, buckwheat and other whole grains. Be sure to *chew* your raw salads thoroughly for greater enzymatic vigor. Enjoy a protein food, too. Dessert should be a raw plant food.

Dinner: Enjoy a protein food in combination with a large plate of raw fruit or vegetables. Enzymes in the raw plant foods will also guard against putrefaction on the intestines and thereby boost assimilation and help control weight.

Snacks: Have a bowl of seeds, nuts, fresh fruit or vegetables for staisfying the snacking urge. These enzymes foods will also act as catalysts in your system to help alert the metabolic process that helps guard against weight buildup.

Sample Enzyme–Catalyst Diet Patterns

Breakfast	Luncheon	Dinner
Citrus fruit Fish, sardines, or cottage cheese, or egg, or cereal with skim milk Whole grain bread Beverage	Sandwich or salad of seafood, fish, cottage cheese Raw vegetable Raw fruit Beverage	Soup; fruit, tomato juice Poultry, meat or fish Potato, brown rice or noodles Cooked vegetable Raw vegetable salad Gelatin, sherbet, fruit Beverage

Snack Suggestions	Dessert Suggestions

Fruit, fresh or sun-dried
Nuts—walnuts, pecans,
   almonds, peanuts
Melba toast with peanut
   butter or fruit puree
Skim milk
Cottage cheese with fresh
   fruit or vegetable slices
Raw vegetables
	Fresh fruit
	Gelatin
	Sherbet
	Almond macaroons
	Angel or sponge cake
	Cake and pastry made with
	   vegetable oil
	Fresh fruit salads or compotes

Restrictions:
Avoid: Cakes made from saturated fats, pastries, doughnuts, muffins, chocolates. If you *must* eat these foods, then you *must* begin and end with fresh citrus fruits for stronger enzyme catalyst action.

THE EVERYDAY ENZYME FOOD THAT CATALYZES
FAT FOR PERMANENT WEIGHT LOSS

Joan G. loved to eat heavy, fatty foods. She always claimed she was "born fat." This was her excuse. While she was born with a certain amount of adipose or fat cells, she could not claim she was born with fat-clogged cells. Rather, Joan G. was fat because she loved to eat those foods that build fat in the cells.

When Joan G. tipped the scales at 244, she decided something had to be done. In addition to the heavy poundage, she had a heavy waistline and embarrassing backside. She felt the burden of her weight whenever she had to get up from a chair. How could she continue eating her favorite fat foods and slim down? She was told about the catalyzing power found in one everyday fruit.

Grapefruit: The Delicious Fat-Melting Fruit. Joan G. was introduced to the tasty golden grapefruit. Available just about everywhere, throughout the year, for a small cost, it is Nature's powerhouse of enzymes that work to catalyze fat and reduce the amount of carbo-calorie fat that add weight to the billions of body cells and tissues.

Grapefruit Vs. Bread. Joan G. had tried so-called "reducing" bread but it did not live up to its promises. The grapefruit did work its miracle weight loss. Here's how. In order to digest a slice of "reducing" bread, it absorbs five times as much of its own weight in water or body fluid. Bread or any other refined food may be a weight gainer because it does not permit the tissues to excrete fluids from their fat-logged cells.

On the other hand, *grapefruit enzymes, in the process of catalysm, acts as a blotter or sponge soaking out cellular fats for excretion through the body's eliminative channels.* Fatty wastes are, in a sense, watery wastes. In order to burn up body fat, grapefruit enzymes are excellent in creating this catalyst-action, as compared to the so-called "slimming" action of "reducing" bread which acts as a sponge.

Low-Calorie, High-Enzyme. One-half grapefruit, contains a low 45 calories. It contains very high enzyme power which makes it effective for fighting fat buildup in the adipose tissues. The enzymes also build up the alkaline reserve of the blood which helps to neutralize accumulated fats and render them more soluble so they can be disposed of. The grapefruit sends a supply of needed enzymes throughout the body to catalyze fat accumulated in the adipose tissues, and cleanse the body of accumulated weight-building elements. It is Nature's own catalyst for fast, permanent weight loss.

Simple Menu Change Melts 97 Pounds. Joan G. made a simple menu adjustment. Before any meal, she would eat one-half or a complete grapefruit. If she tired of the grapefruit, she would try a glass or two of fresh grapefruit juice. Or, she would make a fruit compote of grapefruit wedges with other fruits, and have a sprinkle of honey on top for a sweet taste. Other times, she would make a grapefruit salad of assorted seasonal fruit slices. But always, the grapefruit would predominate.

Results? In a short time, she had melted away 97 pounds. She continued eating her favorite foods although she noticed less of a compulsion for fatty meats. The grapefruit enzymes had adjusted her appestat and she found freedom from the lust or craving for fatty foods. Soon, she was down to a slim 138. Her waist was slim, her backside was trim. She could get up and sit down with the agility of a youngster. She had a lovely silhouette figure. But most important, she could enjoy most of her favorite foods. All she had to do was to include the high-enzyme grapefruit *before* eating any heavy or fatty food. It was the tasty and all-natural way to enjoy good food . . . and keep slim . . . *permanently.*

HINTS AND TIPS TO ENJOY MENU PLANNING ON THE ECD PROGRAM

Meat: Portions should be moderate, with as much fat trimmed off as possible. Broil or bake, rather than fry, for better digestion. Do not add flour when cooking meat as this makes it tougher for enzymatic digestion.

Fruits: Emphasize fresh fruits as much as possible. If unavailable, try water-packed fruits or frozen fruits. Avoid those to which sugar has been added as the enzymatic content will be depleted by this refined sweetening.

Vegetables: These should be raw, if possible. Vegetables that must be cooked should be done so very rapidly in a small amount of water to help preserve as much enzymatic power as possible. Use vegetables, raw or cooked, daily.

Eggs: It's usually best to prepare eggs by soft- or hard-cooking or by poaching. Enzymes can better metabolize egg fat when cooked in these methods.

Fats: Use margarine in place of butter, if preferred. The margarine should be made from liquid vegetable oil and so stated on the label. Enzymes favor polyunsaturated vegetable oils for better mobility in the digestive system.

Beverages: Drink fresh fruit and vegetable juices often. Otherwise, you may prefer coffee substitutes such as Postum, Ovaltine, or herbal teas. Most supermarkets and health stores have these beverages. Enzymes work harmoniously in a fresh juice environment, so emphasize these as often as possible.

Salad Dressings: Use high-enzyme lemon or lime juice or any fresh fruit juice. If you prefer, use the low-calorie salad dressings available. Your enzymes will be invigorated if treated to a salad dressing made of apple cider vinegar and polyunsaturated vegetable oil.

Bread: Select whole grain breads which contains vitamins and minerals and grain protein which propel enzymes throughout the system. Be moderate in your use of bread since they tend to absorb much needed body liquids and may "dehydrate" enzymes. If you want to eat much bread, then increase intake of fruit and vegetable juices to boost enzyme intake.

Nuts, seeds. These should be eaten raw. Chew them very, very thoroughly. The chewing action is most beneficial since it alerts sluggish digestive enzymes to boost their activity. After a *light* meal, a dessert of assorted nuts and seeds can do much to boost digestion and help melt away accumulated fat clinging stubbornly to tissues.

Simple, everyday foods appearing on your menu can add delicious taste to your meals and also give you the enzyme-catalyst foods needed to take weight off . . . and keep it off—while you keep right on eating most of your favorite foods!

Summary:

1. Feature enzyme foods in menu planning. It's easy with the simple 5-step Menu-Planning Guide.

2. Edward B. followed a simple program and lost 14 inches from his formerly fat 48-inch waistline. The fat stayed off *permanently.*

3. Enjoy good foods on the ECD meal planning program, in just six easy steps.

4. Plan your three (or more) daily meals with enzyme foods and enjoy eating while taking off weight.

5. Joan G. lost over 100 pounds, while indulging in her favorite fat foods, but eating a high-enzyme golden fruit, with each meal.

6. It's easy, it's fun, it's delicious to menu plan on the Enzyme-Catalyst Diet Program. It's the delicious way to lose weight.

13

How to Buy, Prepare, Cook Foods For ECD "Forever Slim" Health

Your neighborhood food market has most of the items you need to shed pounds on your Enzyme-Catalyst Diet Program. You need to select tasty foods that boost your enzymatic supply so they can help break down, metabolize and dissolve excess amounts of carbohydrates, calories and fats. One basic rule of thumb when buying, preparing or cooking foods is that *the more natural, the less processed and the least cooked is the stronger enzyme source.*

To give your fat cells the needed enzymes for slimming down, here are basic suggestions:

FRUITS AND VEGETABLES

How To Buy: Locally grown fresh fruits and vegetables are stronger in enzymes because of less storage and shipping. Select plant foods that have a very rich color, with as few bruises as possible. If you must buy canned or frozen, select the variety with as few additives as possible. Many will say the product has no additives. Canned fruits should be the "water-packed" variety so that they have no added sugar which is antagonistic to the enzymatic system and also adds calories. Frozen or canned vegetables should be *without*

174

prepared sauces which are high in calories, fats and also anti-enzyme additives.

How To Prepare: If the fruits and vegetables require no cooking, then eat them raw for strong enzymatic reaction. Eat them as desserts or in salads. You may also eat these plant foods as meals, in themselves, to fortify your system with adequate enzymes that will help catalyze accumulated carbohydrates, calories and fats in the adipose cell tissues. Eat plant foods daily.

How To Cook: Some vegetables do require cooking. Do so in as little water as possible, just long enough to make tender. Eat as soon after cooking as possible. This helps preserve many of the enzymes. Since all fruits can be eaten raw, it is wise to avoid cooking them for maximum enzyme content.

DAIRY PRODUCTS

How To Buy: These are good sources of minerals which are used by enzymes to help maintain body water balance and to help "wash" away fat-causing elements. Skim milk should be used for lower calorie intake. Fermented milks such as buttermilk and yogurt are helpful because they contain helpful organisms which work with the enzymes to digest accumulated substances in the fat cells and help control weight. If you find skim milk too watery, use the "99% fat-free milk" which has 1% fat instead of the usual 3.5 or 4%. These products do not overburden the enzymatic system or make it sluggish. Cheeses are healthy fermented foods which offer minerals as well as friendly organisms to the digestive system for youthful assimilation and metabolism of accumulated pound-building elements. Select skim milk cheeses for fewer calories so your enzymes are not overworked.

How To Prepare: Add fresh fruits or vegetables to many dairy products for healthful vita-mineral enzymatic harmony. This will provide your metabolism with the necessary ingredients for healthful catalystic action and slimming down.

How To Cook: Wherever possible, eat dairy products without cooking since this will weaken some of the vitamin and mineral content and may also destroy some of the needed natural ferments which boost enzymatic power. If you must cook, just steam lightly and then eat as soon after preparation as possible.

MEATS

How To Buy: Whenever you have to buy meats, select lean, well-trimmed and non-marbelized cuts—Grade A is preferable to prime or choice. The lean cuts of beef are those with more muscle than fat, such as round, rump and tenderloin. Avoid cuts such as rib roast and steaks in which the fat is distributed throughout and cannot be removed. Once you have selected your meat, have the butcher trim away all visible fat. (At home, cut off and discard any he may have missed.) Have hamburger ground to order from lean round or lean chuck. Do not buy it already ground unless it is in packages especially marked *lean* ground meat. This is sold by some markets at a slightly higher price. Veal cutlets and chops and veal steaks are good choices, because they are lean. So are veal roasts. *Do not buy canned or frozen cooked meat dishes. You have no way of knowing how much fat, or what kind of fat, they contain.* Remember that fish, chicken, turkey and cornish hens are good because they contain less fat than most meats.

How To Prepare: All visible fats should be trimmed off. You may bake, broil, stew in cooking. Use a rack to broil or roast your meat. Feature broiled meat, if possible, even in mixed dishes and stews. *After cooking a stew, refrigerate, spoon off the congealed fat and heat it again to serve.* You can enjoy your favorite meats on the ECD slimming program. But you have to remember that meats are high in calories and fats. This means that you need to provide strong enzymes to metabolize these weight builders. *The simple suggestion is to eat fresh, raw fruits whenever you eat meats.* The fruit enzymes will attack the tough, connective tissues of the meats and help metabolize them, dissolve the excess fat and prevent their excessive storage in the adipose tissues. Eat raw fruits together with meat at the same meal for the catalyst action that helps keep you "forever slim" while you enjoy your favorite meats.

How To Cook: Try not to "drown" or "choke" enzymes with improper cooking methods. The meat you use will be tasty, and lean. Use a rack when broiling, roasting or baking so that the fat can drain off. If possible, *do not baste,* since basting returns some of the fat to the food and coats it, making it more difficult for enzymes to digest it. (To add more enzymatic power, pour fresh fruit or vegetable juice over meats while cooking or just before serving.) *When you make stews, boiled meat, soup stock or other dishes in which fat cooks out into the liquid, do your cooking a day ahead of time. After the food has been refrigerated, the hardened fat can be removed easily from the top.* Avoid pan-frying meats since this creates a "hard coat" that wears out and depletes enzymes and leads to more cellular fat buildup. If a recipe calls for browning meat before combining it with other ingredients, *try browning it under the broiler,* instead of in a pan. This helps reduce saturated fat intake so enzymes can work with less heavy interference.

EGGS

How To Buy: Buy eggs from a store that keeps eggs in refrigerated display cases. Heat lowers egg quality rapidly and renders them less digestible and in need of stronger enzyme action. Shell color depends upon the breed of the hen and has no influence on the grade, food value, flavor or cooking performance of the egg.

How To Prepare: At home, refrigerate eggs promptly. Store with large end up for better distribution. Before cooking any egg, let it remain at room temperature for a few moments. This makes it easier for enzymatic metabolism.

How To Cook: Healthful methods include baking, soft boiling, and in a variety of dishes that can be broiled. Avoid top-of-the-stove cooking since this congeals the yolk and requires strong enzymatic action for assimilation. Eggs are not too high in calories, but are high in cholesterol (250 to 300 milligrams per egg) and require more enzymes to guard against fat buildup in cells. A combination of eggs with fresh fruit can help metabolize the cholesterol and fat. About three eggs per week can be enjoyed in this manner.

WHOLE GRAIN PRODUCTS

How to Buy: Whole grain breads and cereals as well as such products as brown rice, wheat spaghetti and macaroni, can be enjoyed. Select those that are unbleached and without additives.

How To Prepare: Non-processed whole grain products such as wheat germ, granola type cereals, are prime sources of enzymes so should be used uncooked. Add skim milk, some fresh fruit slices, and you have an enzyme dish that is as delicious as it is healthful. For brown rice, spaghetti and macaroni, eat with a topping of raw vegetables to metabolize the carbo-caloric intake and guard against adipose tissue buildup.

How To Cook: If you must cook whole grain foods, do so with as little water as possible, in a covered kettle to help seal in as many nutrients as possible. Serve whole grain foods with fresh fruits or fresh fruit juices for better enzymatic assimilation.

FISH

How To Buy: Eat fish throughout the year. Avoid shellfish which come from close-to-the-shore polluted waters and are destructive to enzymes. Select any fresh or salt water fish for tasty eating. Fish is high in polyunsaturated fats which are used by enzymes to help break down accumulated fats in the adipose cells. Fresh fish is best. Frozen or water-packed canned fish can also be used in the absence or unavailability of fresh fish.

How To Prepare: Favorable methods include baking, broiling, stewing. Avoid frying since the fish becomes coated with a thick envelope that is difficult for enzymes to penetrate.

How To Cook: Use any polyunsaturated vegetable oil. Top with a squeeze of fresh lemon or lime juice for greater enzymatic penetration. Keep breading to a minimum since this coats the fish with a thick layer that forces enzymes to work harder.

BEANS, PEAS, LENTILS

How To Buy: These are excellent sources of protein and should be featured often. Enzymes use these plant proteins for strong cellular washing. Buy dried beans, peas, lentils. Store in a glass jar in a cool section of your pantry. Plan to use within six months after purchase.

How To Prepare: Soak overnight in water, remembering that they will double in length after the liquid is restored. Then cook the next day in the same liquid.

How To Cook: Bring to a boil, then reduce to a simmer and let cook, covered, until soft enough to eat.

How A "Secret Enzyme Food" Helped Create "Forever Slim" Health. Ellen X. could not dispose of some 38 stubborn extra pounds. Her breathing was labored. She tired easily. She was constantly letting out her clothes as more and more weight caused her to spread out in all directions. She loved to eat a variety of beans and peas as well as lentils. She found a "secret enzyme food" that would turn these ordinary foods into dynamic fat-melters. She would soak beans overnight in *fresh fruit juice!* Next morning, she would cook the beans in the juice. **Benefits:** Enzymes in the fruit juice boosted the strong enzyme power of the beans. When she ate the beans, she fed her system a double-barrelled supply of fat-melting enzymes. Soon, she lost close to 35 pounds. Her figure was trimmer. She breathed better. She had more energy. When she lost over 40 pounds, she told other "fatties" of her "secret enzyme food" and boasted that she had lost weight without exercising, attending clubs, without starving . . . but she lost weight by eating!

HOW TO ENJOY FATS AND OILS ON THE ECD PROGRAM

No need to give up your love for the taste of fat. Just change to polyunsaturated vegetable oils that give you the same taste-satisfaction as so-called hard fats. These oils are needed by enzymes to be carried in an emulsion like environment, throughout your circulatory system to help wash away accumulated fats from your cells. Enzymes can "dry out" or "dehydrate" without this unctuous environment, so fats and oils can help you keep slim on your Enzyme-Catalyst Diet Program.

Select polyunsaturated vegetable oils available at almost all markets. As a start, try using oil in these ways:

1. In broiling or baking fish or poultry.

2. As an ingredient in barbecue sauces and sauces for marinating meat, chicken, or fish. (When you broil cubes of beef or lamb that have marinated for several hours in an oil-vinegar-seasoning mixture, you have tasty shish kebab.)

3. In French dressing and mayonnaise.

4. In cooking vegetables with little or no water. (Use 1 teaspon of oil for each serving. Put oil and vegetables in skillet with tight cover, season. Cook over very low heat, stirring occasionally, until vegetables are done—about 5 to 15 minutes. Add a little water or other liquid during cooking if needed.) This helps boost and retain good enzyme vigor.

5. As a seasoning, with herbs and apple cider vinegar or lemon juice if desired, for cooked vegetables.

6. In browning meats for stews, pot roasts, etc.

7. In oven-frying meat, fish, poultry and some vegetables.

8. To saute onions for onion soup (or vegetables for vegetable soup). Use fat-free bouillon or consomme as stock.

9. In cream sauces and cream soups made with skim milk.

10. In whipped or scalloped potatoes, with skim milk added.

11. For making hot breads such as biscuits, muffins, etc.

12. In home-baked beans or in casseroles made with dried peas or beans.

Finally, create a high-enzyme salad with polyunsaturated oil and fruit juice poured in a bowl containing raw vegetables and a sprinkle of wheat germ and a bit of honey. It's a delicious way to stimulate enzymatic vigor and help guard against cellular obesity.

FOUR WAYS TO PROTECT ENZYME ACTIVITY

Enzymes can be inhibited, slowed down and even destroyed if you eat corrosive, chemicalized foods. In particular, an excessive intake of hard fats may immobilize enzymes. This means that the fatty foods you eat will be deposited in your adipose cells, then cause weight buildup. Enzymes will be nullified by excessive hard fats. Here are four ways to protect your enzyme activity when selecting foods:

1. *Packaged Foods.* You can enjoy any kind of prepared or packaged food that contains *no fat at all.* Examples are vegetarian baked beans and angel food cake mix. Your enzymes will be spared destruction if the packaged food is free of additives.

2. *Fat In Foods.* You can enjoy prepared or packaged foods if they have fat that comes from a polyunsaturated source. Examples are sardines packed in soybean or vegetable oil. Your enzymes can use these oils for better mobility and smoother fat-melting activities.

3. *Restrict Frozen Dinners.* Do not buy frozen dinners or other ready-to-eat canned or frozen food mixtures which contain fat. You usually cannot tell what kind of fat was used or how much. Your enzymes may become choked if there is hard fat in the frozen food and your cells may become clogged with excessive deposits.

4. *Fat-Free Prepared Foods.* Dehydrated foods, such as potatoes and mixes to which you add the oil yourself, are satisfactory, if you must eat them. Examples are pancake mixes. Read the label to be sure they do not contain any fat. By adding your own polyunsaturated oils, you are giving your digestive system needed lubrication upon which enzymes can slide in all directions for better catalyst action.

With some simple adjustments, you can enjoy good taste, good food and good enzymatic catalyst action to help keep you "forever slim."

Main Points:

1. Your neighborhood food market is a good source of almost all enzyme foods you need to help you catalyze your accumulated carbohydrates, calories and fats.

2. For stronger ECD power, note how to buy, prepare and cook dairy foods, meats, eggs, whole grain products, fish, beans, peas, lentils.

3. A "Secret Enzyme Food" helped Ellen X. shed more than 38 stubborn pounds so she could enjoy "Forever Slim" health.

4. Slim down while you enjoy fats and oils on your ECD Program.

5. Protect enzyme activity in four easy ways.

14

How to Keep Slim
While Dining Out
on the ECD Program

Whenever Arthur O'L. dines out, he scans the menu for *one* group of items that he must have if he wants to keep slim on the Enzyme-Catalyst Diet Program. The group consists of *raw* foods. Arthur O'L. discovered this secret from a fellow salesman when he envied the other man's slim shape. Arthur O'L. was some 39 pounds overweight and still gaining. Even when he dieted, Arthur O'L. could not shed many of the corpulent 39 pounds. But his slim salesman friend told him that whenever dining out, be sure to order a bowl or a plate of any available raw food—either raw fruits or raw vegetables.

Enjoys Good Foods, Keeps Losing Weight. Arthur O'L. would order any raw fruit plates or raw vegetable salads which he would eat *before* the rest of the meal. He found that not only did this help cut down his appetite, but he enjoyed most of his good and favorite foods, while he kept losing weight. Soon, he lost the heavy hanging paunch; his "double chin" slimmed down. His body was trim and slim as more than 39 pounds were catalyzed away ... while he continued eating.

Enzyme Activity Triggers Fat Melting. When Arthur O'L. ate a raw enzyme food at the start of the meal, the vigorous chewing alerted strong mouth and digestive enzymes so they could work with catalystic power to dissolve the eaten carbohydrates, calories and fats. Even when dining out, this simple trick helped keep Arthur O'L. neat and slim, while enjoying most of his favorite meals.

Party Time Can Be Slim Time, Too. If you attend a party, a festival, an outing or any dinner outside the home, with a little preparation, you can continue on your ECD slimming program.

Simple Slim Tricks. Most of the time, you will be able to do something about the "wrong" foods you are served. You can remove the skin from friend chicken, for example, or the coating from a veal cutlet. With a good sharp knife, you can remove a good deal of the fat from a steak or chop. You can push the gravy aside and eat what's under it.

Remember that the occasional, unavoidable days when you cannot stay strictly on your ECD program are not as important, in the long run, as careful adherence to it for the rest of the time.

BASIC RULE: WHEN YOU EAT OUT

Get into the habit of saying NO to—

Cream soups	Gravies
Fried foods	Cheeses made from whole milk
Casseroles and other	Ice creams
mixed dishes	Puddings
Creamed foods	Cakes, pies and similar
	high-calorie desserts

Choose Freely From These Items, If You Can—

For First Course: clear soup; tomato or any vegetable juice; fruit cup.

For Main Dish: fish or chicken (baked or broiled without butter); sliced turkey or veal; London broil (flank steak) without gravy; fruit, gelatin or fish salad (with dressing or mayonnaise served separately); vegetable plate (if butter has not been added).

For Dessert: as much fruit as possible; sherbet; gelatin; unfrosted angel food cake.

HOW TO CONTROL OVER-EATING

When you dine out, the temptation to over-eat is great. You don't have to wash any dishes so you can continue eating. This can cause weight buildup. Behavior modification consists of substituting good eating processes for poor ones to keep your enzymes functioning adequately. Typical behavior modification plans are:

- Eat only when sitting down, no matter where you are. This helps you plan your eating and puts a natural control on your appetite.
- When you eat, do just that. No activities, no watching television, no reading with meals. This puts a control on "mindless eating."
- Reward yourself a little. While eating, keep telling yourself that your enzyme system will digest and metabolize the foods. This creates a mental influence that boosts the catalytic power of enzymes and improves digestion.

12 WAYS TO KEEP SLIM WHILE DINING OUT

Enjoy good taste with most foods offered to you when you have to dine away from home, but keep these points in mind:

1. Try not to fill yourself up to the brim. This can overwork your enzymes and add more weight to your fat cells. Be good to your enzymes by eating smaller portions of foods, chewing very thoroughly, satisfying your oral urge. This helps control a runaway appetite.

2. Pass up sugary or sweetened beverages. Instead, ask for fresh fruit or vegetable juice. Nearly every restaurant or party place will have such juice. This is your enzyme fortification method that helps build up digestive catalysts in reserve to dissolve accumulated fats to be eaten later.

3. Steer clear of any rich gravies or sauces, heavy dressings. Keep your enzymes "clean" of such heavy sauces.

4. Many places offer tempting pastries, rich, creamy custards and desserts which are laden with sugar that can drain out the energy of your enzymes. Pass these up. Instead, seek out gelatin desserts, fresh fruit desserts or even assorted seeds and nuts. These give you an abundance of enzymes while helping to keep you slim.

5. If you feel very tempted to partake of more food than you should, resist the urge to drink more fruit or vegetable juices, or else try coffee or tea without any sugar.

6. Keep away from cold soups, especially if they are creamed. They are lifeless and also coat the digestive tract and inhibit

the activity of your enzymes. Select comfortably warm vegetable soups.

7. Soft, cold foods are usually high in fats or calories or both. You should know what the food is made out of, before you eat it. You're always safe with a platter of fruits, vegetables, cheese wedges, even cold roast beef slices.

8. Cooked foods should have fat skimmed off. Do this with your spoon. This will help your enzymes work effectively without the burden of excessive fats.

9. Tell yourself that you will eat only if you are hungry. If you force yourself to eat, you upset your enzymatic harmony and you tend to build up excessive weight that way.

10. Always select fresh raw fruits and vegetables first. If they make up the appetizer, you'll cut down on the tendency to overeat. More important, you'll be sending a fresh supply of enzymes to cope with the food that will later be eaten.

11. The catalyst power of enzymes is strong when you eat foods that are baked, boiled or broiled. Select these foods from any available list or table.

12. Raw fruits and vegetables are often available in convenient-to-eat form. If you prefer, bring along a supply for munching. For example, you can bring a banana, radishes, chopped celery, cucumber slices, mushrooms, green peppers, watercress, carrot strips, water chestnuts. Just have them in a small container for munching before any meal. Any seasonal fruits, such as apples, pears, berries, etc., are also good enzyme foods that bolster the fat-melting powers in your system.

With some few adjustments, you can enjoy eating out while still fortifying your metabolism with enzymes that help keep your adipose cells reasonably free of excess fats.

YOUR DINING OUT HIGH-ENZYME, LOW-CALORIE MENUS

Here are sample menus that are high in enzymes and low in calories. You can order these foods from almost any menu when you dine out:

1200 CALORIE MENU #1

Breakfast

Chilled half grapefruit
3/4 cup dry cereal
1 cup skim milk
1 soft-boiled egg (optional)
1 slice whole grain toast
1 teaspoon polyunsaturated margarine
Coffee or tea

Luncheon

Tomato stuffed with chicken
(use 1 tomato; ½ cup diced chicken;
2 teaspoons mayonnaise made from
low-fat ingredients; capers, parsley;
celery; lettuce)
1 small hard roll
1 cup skim milk
1 small banana, sliced
Coffee or tea

Dinner

Baked fish fillet (4 ounces)
with 1 teaspoon oil
Broccoli with 1½ teaspoons Hollandaise sauce
Scalloped tomatoes (use ½ cup
sliced tomatoes; 1 slice diced whole grain
bread; 1 teaspoon oil; basil)
1 fresh pear
Coffee or tea

1200 CALORIE MENU #2

Breakfast

½ cup orange juice
2 plain whole grain muffins
Coffee or tea

Luncheon

Open roast beef sandwich (use 2 thin slices–
3 ounces–lean roast beef; 1 slice bread)
½ cup cole slaw with 2 teaspoons low-fat mayonnaise
Sliced sour pickles
Melon or other fresh fruit in season

1 cup skim milk
Coffee or tea

Dinner

¼ barbecued chicken–3 ounces–with
    2 teaspoons oil)
1 small ear corn
½ cup mixed carrots and peas
½ cup fresh fruit cup
Coffee or tea

1800 CALORIE MENU #1

Breakfast

Chilled half grapefruit
¾ cup dry cereal
1 cup skim milk
1 soft-boiled egg (optional)
1 slice toast
1 teaspoon polyunsaturated margarine
1 teaspoon marmalade
2 teaspoons honey for cereal, fruit or beverage
Coffee or tea

Luncheon

Tomato stuffed with chicken (use 1 tomato;
    ½ cup diced chicken; 2 tablespoons low-fat
    mayonnaise; capers; parsley; celery; lettuce.)
1 large or 2 small hard rolls
1 teaspoon polyunsaturated margarine
1 cup skim milk
1 small banana, sliced
Coffee or tea

Dinner

Baked fish fillet (4 ounces); use
    1 teaspoon oil and 1/4 cup whole grain bread crumbs
Broccoli with 1½ teaspoons Hollandaise sauce
Scalloped tomatoes (use fresh tomato slices;
    1 cup diced bread; 1 teaspoon oil; basil)
1 baked potato
1 teaspoon polyunsaturated margarine
1 pear
Coffee or tea

1800 CALORIE MENU #2

Breakfast

> 1/2 cup orange juice
> 2 plain whole grain muffins
> 1 teaspoon polyunsaturated margarine
> 1 cup skim milk
> Coffee or tea

Luncheon

> Roast beef sandwich (use 2 thin slices or
>   3 ounces, lean roast beef; 2 slices whole
>   grain bread)
> 1/2 cup cole slaw with 1 tablespoon low-fat mayonnaise
> Sliced sour pickles
> Melon or other fresh fruit in season
> 1 cup skim milk
> Coffee or tea

Dinner

> 1/2 cup chilled fruit juice
> 1/4 barbecued chicken, 3 ounces with
>   2 teaspoons oil
> 2 small ears corn, or 1 large ear, with 2
>   teaspoons polyunsaturated margarine
> 1/2 cup mixed carrots and peas
> Mixed raw green salad
> 1 tablespoon low-calorie French dressing
> 1 slice French bread
> 1/2 cup sherbet
> 1 cup skim milk
> Coffee or tea

HOW TO CHEW YOUR WAY TO A NEW SLIM SHAPE

Mabel and Fred P. were a "fat couple" until they followed a simple enzyme method of boosting the fat-catalyzing power of these substances, even when dining out as they often did.

Seeds, Nuts, Chopped Vegetables. The "fat couple" carried a supply of seeds, nuts and chopped vegetables in a glass jar, whenever they had to attend a social function that called for a lot of eating. Before any food, they would eat several handfuls of seeds, nuts and chopped vegetables. *They would chew very thoroughly.* This alerted digestive enzymes which were carried by the polyunsaturated oils of the chewed seeds, and nuts, and transported by the minerals of the vegetables, to all parts of the body. Foods eaten afterward were catalyzed by the "waiting" enzymes and there was less of a weight buildup. The "fat couple" soon slimmed down and were known as the "slim couple" who could eat at social affairs but still keep slim.

THE ENZYME APPETIZER THAT CONTROLS APPETITE

Laura M. is a formerly-fat caterer. Previously, it had been tempting for her to taste and sample foods at a banquet or buffet or party which she arranged. She started to "bloat" up because of the excess food. So she discovered a simple enzyme catalyst method that helps metabolize foods, but more important, *it controls her appetite.*

Before-Meal Appetizer. Eat a fresh raw fruit or vegetable or a fruit or vegetable salad about 30 minutes *before* you will be tempted by the "groaning table" at a social gathering.

ECD Benefit: Enzymes will take the carbo-calories from the fruit or vegetable and send it into the bloodstream, then directly to the brain's appetite control center. Here, the enzymes use the substances in the raw plant food to establish a stopgap type of control to block off the nerve impulse to eat compulsively. This is the enzyme's way of appeasing your appetite and helping you ease the urge to eat.

Loses Weight, Controls Appetite. Laura M. has been able to enzyme-melt some 35 unwanted pounds. Furthermore, she is *permanently* slim even when faced with tempting food because she uses the Before-Meal Appetizer regularly. Simple, yet slimmingly effective.

You can enjoy many of your favorite foods when dining out and still keep slim with the power of raw food enzymes available almost on any menu or at any social gathering.

In Review:

1. Arthur O'L. lost over 39 pounds even when dining out, on the simple ECD Program.
2. A few foods are taboo; most are acceptable on the ECD Program when dining out. check the list.
3. Note the 12 ways to keep slim, enjoying good, tasteful food, while dining out.

4. Enjoy the 1200 or 1800 high-enzyme, low-calorie menu plan when dining out.

5. Mabel and Fred P. went from "fat couple" to "slim couple" by chewing foods thoroughly.

6. Laura M. melted 35 unwanted pounds and put a natural control on her appetite at social dinners with a simple Before-Meal Appetizer.

15

How Enzymes Soothe "Runaway Gland" Eating Urges

Your body is a network of glands that can create a compulsive eating urge with such effects that you put on dozens and dozens of pounds. These glands need adequate enzymatic nourishment so they can self-regulate the amount of hormones they secrete and help soothe a "runaway" eating urge. Let's take a closer look at these glands and see how the Enzyme-Catalyst Diet Program can satisfy them and control their secretions and thereby control the eating urge so you can take weight off permanently.

Glands Influence Your Weight. Basically, a gland is a small sac-like body organ or structure which manufactures a liquid product from its cells. Most glands secrete their liquids through various channels or ducts. The *endocrine* glands are those which do not excrete their liquids but leave it to be picked up and distributed by your bloodstream to other parts of your body. These hormones are used by enzymes to create a dissolution of accumulated carbohydrates, calories and fats from your adipose tissues. The hormones are activated by enzymes to dislodge accumulated weight-causing elements from the adipose tissues, to metabolize them, to help cast them out of the body. This hormone-enzyme reaction helps control your weight. This biological process works in smooth rhythm to help keep your weight off, *permanently.*

THYROID GLAND: ENZYMATIC KEY TO
PERMANENT WEIGHT LOSS

Location: A two-part endocrine gland that looks like a butterfly, it rests against the front and on either side of the windpipe.

Hormone: Secretes the *thyroxin* hormone which stimulates the activity or metabolism of adipose fat cells and tissues.

How Enzymes Influence Thyroid Gland. Raw food enzymes take myglobin from iron-rich foods and blend it with iodine to act as a place of storage for oxygen in the muscle tissues. Enzymes then use this iron-iodine mixture to activate the function of the thyroid to metabolize carbohydrates, calories and fats and dissolve them from the adipose cells and thereby help to create weight loss and a slimmer figure. *A well-regulated and enzyme-nourished thyroid gland works with biological rhythm and can create a permanent weight loss.*

How A Natural "Enzyme-Hormone Tonic" Melted Weight Forever. Shirley McB. lamented over her excess weight. She tipped the scales at 202 pounds! Furthermore, she had an unusual sensitivity to cold. She was physically weak and lethargic. She also developed dry, coarse pale skin; her once lovely blonde hair became dry and brittle. She had very low metabolism and a "sleepy" thyroid gland. A low *thyroxin* meant that the sluggish enzymes could not metabolize the excess foods deposited in her billions of adipose or fat cells and tissues. Furthermore, the weight gain was accompanied by the overall sluggish symptoms that accompany a condition of low thyroid and weak enzyme activity.

Shirley McB. had tried various reducing programs but she still felt sluggish, weak and kept putting on weight. She recognized that her fault was in treating symptoms of overweight, rather than the *cause.* When she was told that she had a low iron-iodine intake, that her enzymes needed these minerals for better fat metabolism, she began to boost her nutritional needs.

Iron + Iodine = Enzyme Catalyst Action. Shirley McB. prepared a simple "Enzyme-Hormone Tonic." In a glass of fresh vegetable juice, she stirred one teaspoon of wheat germ, one teaspoon of blackstrap molasses, one-half teaspoon of sea salt, one tablespoon of cod liver oil. She would stir vigorously. She would drink one glass in the morning. A second glass at Noontime. A third glass as a nightcap.

Benefits: The iron in the wheat germ and molasses were energized by the iodine in the sea salt and cod liver oil. The enzymes in the vegetable juice took this iron-iodine combination, nourished the thyroid gland, stimulated this sluggish organ to secrete a required amount of *thyroxin* which was then alerted to create better cellular metabolism. The "Enzyme-Hormone Tonic" used enzymes to *adjust* her hormone clocks so that the adipose tissues could be cleansed and "reduced" of accumulated carbohydrates, calories and fats.

Loses 74 Pounds, Feels Warm, Energetic. Shirley McB. began to melt pound after pound. It was a joy to step on the scale and see the weight go down. More benefits of the enzyme-stimulated glandular adjustment were feelings of comfortable warmth, better energy; her skin glows with the look of "peaches and cream" and her blonde hair began to take on a natural gleam. When Shirley McB. slimmed to a neat 128, she felt alive with the joy of youth. Now she takes just one "Enzyme-Hormone Tonic" each day to keep her thyroid gland nourished, to control her appetite and to keep herself looking and feeling permanently slim! (All ingredients are available at most food stores or special health food shops.)

HOW A 5¢ NATURAL HORMONE FOOD CREATES PERMANENT WEIGHT LOSS

Allan K. is a compulsive eater. He admits that he has a "runaway appetite." Whenever he sees food, he has the insatiable urge to eat and eat and eat. When eating in restaurants, he orders the biggest meals available. Rarely is he satisfied with less than two heavy, sweet desserts. To add to this distressful problem, an hour after he has eaten, he is looking for "snacks" or "nibbles" that often turn into small meals, in themselves. Allan K. is a victim of a "runaway gland" problem. His thyroid gland is sluggish. A *thyroxin* deficiency has given him an uncontrollable appetite. He knows this but is afraid to take a synthetic drug because of reactions.

Fish Liver Oil Acts As Natural Hormone. Allan K. wanted to feed his thyroid gland with a natural hormone. He found out that fish liver oil is a prime source of ocean minerals, especially *iodine.* He started to take two tablespoons of any fish liver oil (cod liver oil, halibut liver oil are two good iodine sources, available at almost all pharmacies or health food stores) in a glass of fresh vegetable juice. Allan K. would take this natural hormone food three times daily. He estimated its total cost at about 5¢ per glass. The iodine fed his sluggish thyroid and promoted a healthier secretion of needed *thyroxin.*

Enzymes Regulate Hormone Balance, Control Appetite. The enzymes in the vegetable juice took the iodine from the fish liver oil, alerted the sluggish body metabolism, helped to regulate a healthful hormone balance. Once the bloodstream was fed a steady supply of enzyme-propelled iodine, Allan K. felt less of a compulsion to keep on eating and eating and eating. Soon, Allan K. lost some 68 pounds. His waistline was trimmed down to a neat 34 and it was still slimming down. He felt better and lighter. When he used this enzyme-iodine combination, his metabolism was alerted so that his gland could be soothed, even pampered, so there was no "eating reaction" that could cause overweight.

Now, Allan K. takes just *one* glass of this vegetable juice and fish liver oil tonic daily as a "natural enzyme hormone tonic" and finds that his weight is kept off *permanently.* He enjoys three meals per day, but in moderate and fully satisfying amounts. He eats ... but no longer overeats, thanks to the enzymatic adjustment of his hormone clocks!

ADRENAL GLANDS: NATURAL WEIGHT CONTROLLERS

Location: A small set of glands, shaped like Brazil nuts, they sit astride each kidney. The adrenal glands consist of two parts: the medulla (central portion) and the cortex (covering). Both of these are influenced by enzymatic reaction. Enzymes will stimulate the medulla to secrete *adrenalin* or *epinephrine* (the emergency hormone) in times of emotional stress or danger or when there is a nervous urge to want to eat. Enzymes will use adrenalin to help alert the heartbeat, to send a supply of sugar into the bloodstream, to improve digestion, to send blood into the muscles. Enzymes will also use adrenalin to soothe the cortex so that it can meet the challenges of the situation being faced. A deficiency of enzymes means that the adrenal glands cannot secrete sufficient hormones and the body cannot cope with the challenges faced. In these situations, an enzyme deficiency can weaken the natural desire to control eating. As a consequence, the appetite "runs wild" and there is the compulsive eating urge that puts on weight.

Hormone: Enzymes prompt the adrenal gland to secrete adrenalin which is used for more than just control of the body's instincts, but for better metabolism and digestion of foods. Adrenalin is also used by enzymes to help metabolize accumulated carbohydrates, calories and fats from the billions of adipose fat tissues. Enzymes need adrenalin to help keep the cells slim and thereby control weight.

The Everyday Food That Nourishes The Adrenals To Act As Natural Weight Controllers. Rita J. was admittedly corpulent. She had a heavy waistline, thick arms, thick thighs, protruding derriere. To look at the scales was tortuous because Rita J. was so heavy, bending over caused her to wince and wheeze. When she finally did read her weight, she cried out with shame because she was more than 93 pounds over the safe levels. Her husband, her family, her friends, all whispered and giggled behind her back about her unsightly, comical "fat lady" look. But Rita J. kept saying that she was "always hungry." She would gobble down everything in sight. She was a pathetic sight at the refrigerator, long after others had gone to sleep. She would "raid" the refrigerator and eat just about all that she could, without having to cook it. As for leftovers, they did not stay too long. She would eat them before the night was over.

Citrus Fruits Act As Adrenal Regulators And Natural Weight Controllers. Rita J. had tried chemicalized diet pills but they gave her headaches, nausea, weakness. She needed a natural "diet pill." She was told that her wild appetite was due to her "runaway gland," but she felt she could do nothing about it. Then she was told that enzymes needed a nutrient found in citrus fruits that could be used to nourish her "starved" adrenals and calm them down so that they would not force her to be a compulsive eater. Rita J. followed this simple program:

1. Before breakfast, eat one-half grapefruit.

2. Before luncheon, eat a seasonal melon half.

3. Mid-day, eat a platter of raw citrus fruit wedges, consisting of oranges, lemons, limes, tangerines. Sprinkle with a bit of honey to add a naturally sweet flavor.

4. Before dinner, eat a fresh fruit salad with a low-fat cottage cheese scoop for good mineral intake.

5. As a nightcap, drink a glass of fresh citrus fruit juice. A glass of orange or grapefruit juice or a mixed fruit juice cocktail.

Benefits: Enzymes will use the Vitamin C from the citrus fruit and the minerals from the cottage cheese to stimulate the adrenals to issue hormones which can act as natural tranquilizers for the cerebro-neuromuscular system. This helps soothe and control the nervous urge to eat.

Enzymes will then take Vitamin C and use it for washing the adipose cell tissues and creating collagen, the glue-like substance that will strengthen the cells and protect against overloading with calories or weight-adding substances.

Eases Appetite Urges, Loses 77 Pounds, Slims Down "All Over." Rita J. followed this program for a short time. She noticed, almost from the start, her appetite urges began to subside. Then, she found her weight going down. Weighing herself daily, she could bend over more easily and happily, too, when she saw how the scale kept going down . . . down . . . down. Her heavy waistline, thick arms and thighs and protruding derriere soon slimmed down. She soon counted some 80 pounds melted away. Rita J. felt her greatest accomplishment was the "natural weight control" power of the raw citrus fruits. She had less of a desire to overeat. Her portions were smaller and smaller. Soon, she shed well over 100 pounds and was slim "all over." Her food intake had been cut by one-third. She was *permanently slim* now, because her adrenal glands had become biologically adjusted by the enzyme and Vitamin C power in everyday citrus fruits!

PITUITARY GLAND: MIRACLE REDUCING POWER

Location: About the size of a pea, it is suspended from a slender stalk at the base of the brain. It is often called the "master gland" or "miracle reducer" because its three lobes secrete at least nine known hormones which influence body weight gain and loss.

Hormone: Enzymes alert the pituitary to secrete many hormones, particularly one from the anterior (front) portion which is involved with the thyroid gland to metabolize carbohydrates (sugars and starches.) A weak enzyme supply means a weakened anterior hormone and a buildup of carbohydrates in the adipose tissue cells. Weight gain is known for being caused by a "starved" or "ailing" pituitary gland. Enzymes control the secretions of these needed hormones to metabolize carbohydrates and fats so they do not accumulate in the fat tissues and cause weight gain.

The Appestat—Miracle Reducing Gland. Closely connected to the pituitary in the brain are several portions known as the *hypothalamus.* This segment houses a region known as the "appestat" or appetite-control center of your brain. A very active appestat means you have a nervous urge to overeat. Even a reasonably over-active appestat will cause you to eat more than you should and, untreated, can lead to so-called "impossible" pounds that cannot be taken off. The appestat is the "miracle reducing gland" because when it is controlled through sufficient enzymatic nourishment, it eases the compulsive eating urge. It is the key to effective weight loss and/or gain. It is Nature's miracle reducing gland . . . and you *can* control it through the intake of enzymes in everyday foods.

How Raw Seeds And Nuts Eased "Eating Urge" And Melted 88 "Stubborn" Pounds. Burton S. was a hard working factory supervisor. He always maintained that his heavy work called for heavy food intake. But his problem was *not knowing when to stop eating!* Even if he did not work, he was so habit-trained to overeat, he would devour full plates of heavy foods and then indulge in double and even triple portions of rich, sweet desserts. Burton S. had a 50 inch waistline! He was some 120 pounds overweight! He did not walk. He waddled! When he experienced short breath, when he wheezed and coughed and sputtered, when he had to ask for help in getting up out of a chair or bed, he decided something had to be done. But how could he beat his compulsive eating urge?

He was told that he needed to send a supply of enzymes into his system to nourish his pituitary gland so that hormones would soothe his *hypothalamus* and thereby act as a "natural tranquilizer" for his runaway appetite. But he needed time-release, slow-acting enzymes. That was where ordinary seeds and nuts came into use.

Time-Release, Slow-Acting Enzymes. Raw seeds and nuts are sources of living enzymes. That is, the enzymes are housed within the seeds and nuts, capable of sustaining life and even "giving birth" if need be. Therefore, these are enzymes of the highest biological value since they are self-sufficient and self-sustaining. Furthermore, seeds and nuts are prime sources of polyunsaturated fats and essential fatty acids. When the seeds and nuts are *thoroughly chewed,* mouth enzymes take up the oils, and then "coat" the digestive system with the essential fatty acids. Furthermore, the enzymes of the seeds and nuts float in the oils and create a slow, steady "appetite control" reaction. The seed and nut enzymes are used to soothe the pituitary and other glands. The oils tend to preserve the activity of the enzymes so they work slowly in soothing, pampering and hushing the appestat.

One Handful Controls Day-Long Appetite. Burton S. would eat one handful of assorted seeds and nuts (sold at most food stores, supermarkets, special health stores, specialty shops) in the morning. He found that if he chewed them very thoroughly, his appetite urge became less frantic. He found that for most the day, the "wild craving" was soothed. The "savage hunger" was abated. His portions were halved. Soon, those half-portions were halved. Burton S. melted away some 70 pounds on this easy enzyme program. Gradually, he lost the excess 88 pounds. More important, when he shed close to 120 pounds, *his weight was permanently reduced!* He felt satisfied with normal portions of tasty foods.

Burton S. is able to enjoy his new slim-trim figure by eating one handful of enzyme-high seeds and nuts daily. His appetite is controlled throughout the day. He has become so slim, his co-workers call him "Skinny." He likes the new name!

HOW TO USE PLANT FOODS TO CONTROL EATING URGE AND LOSE WEIGHT

Eat raw fruits and vegetables on an empty stomach! This is the miracle weight losing secret that can shed even the most stubborn of pounds, while putting a natural "stop" to your ruanaway eating urges.

How Enzymes Fight Cellular Fat. When your digestive system is freed of other foods and receives raw fruits and vegetables, it can work without interference to create a unique melting away of cellular fat. Here's how enzymes can keep you slim forever:

Overweight is often traced to an excessive amount of carbohydrates, calories and fats clinging to your adipose cell tissues. The fat is usually present in two very large compartments. The first is the *intracellular.* That is, the fat has invaded the *inside* of your adipose cell tissues. The second is the *extracellular.* That is, the fat has become glued to the *outside* of your adipose cell tissues.

Enzymes Must Go Into Center Of Cell. Your metabolism must send enzymes to the center of your cell. This compartment is called the *interstitial space.* Once the enzymes enter your interstitial spaces, they can then fight the fat from *within* the adipose cells. They can help fight the intracellular fat and then go to the extracellular fat. Enzymes need to go to the cell center. If your metabolism is "busy" with other foods, the power is weakened. Enzymes may not be able to be pushed into the interstitial space. This means weakened cell reducing. Your body glands need enzymes in the interstitial spaces so they can promote better hormone secretion and also help slim down accumulated fats in the gland cells. Enzymes can work best if there is a minimum of interference from having to digest high fat or protein foods. Therefore, your glands can be soothed, your cells reduced, *if you eat raw fruits and vegetables on an empty stomach.*

Enzymes Use Hormones For Cellular Slimming. Raw plant foods send enzymes to your glands, promote secretion of hormones which now exert a molecular reaction. That is, they create *osmotic pressure* within your adipose cell tissues. The cells act as "sponges" in absorbing carbohydrates, calories and fats. But the enzymes use hormones to "squeeze" out these weightly substances from the cells. This is also known as releasing *tissue turgor,* or slimming down the cells. Enzymes use hormones for this slimming action. They also use hormones to cleanse the gland tissues and ease the eating urge at the root cause.

When you eat raw fruits and vegetables ON AN EMPTY STOM-ACH, you send a superior enzymatic power, without interference, throughout your glandular network to control appetite and create this *osmotic pressure* which "squeezes" and "dissolves" carbohydrates, calories and fats from your adipose cell tissues.

It is the natural way to control your appetite and also to help wash away unwanted weight and slim down *permanently*.

Eat Raw Fruits In The Morning. Start your day off right with a bowl of seasonal raw fruits. Use a little honey for sweetening, if desired. Fruit enzymes can work without interference of other foods, to regulate your hormonal system and to "squeeze" fat out of your cells and help you lose weight.

Drink Fresh Fruit Juices For A Meal. Skip a meal for one time and drink several glasses of assorted fresh fruit juices. The enzymes are in a ready-to-use form, having been squeezed from the fruit. They are rapidly assimilated and help create speedy appetite control and cellular reducing.

Eat Raw Vegetables For A Meal. One evening, instead of a customery meal, eat a plate of assorted raw vegetables. For dressing, use apple cider vinegar, oil and a bit of honey. Chew vigorously and thoroughly. This alerts digestive enzymes to work at soothing your glands and also helping to reduce your cells. An entire meal of succulent, juicy, crisp, delicious vegetables offers you exhilarating good taste and a slimmer figure, as well.

A One Day Raw Food Program. Set aside one day a week for raw food. During the day, eat nothing but raw fruits, vegetables, drink juices, eat raw seeds, nuts and grains such as wheat germ, oats. You will be giving your metabolic system *pure enzyme power* to work without interference. If you can have this Raw Food Program several days a week, you'll discover your appetite subsiding, your weight declining, your figure trimming down. Vary the different raw foods and juices for greater taste enjoyment. Soon, you'll discover that you have less and less of a food craving and more and more of a feeling of satisfaction on smaller portions. You'll feel "smaller" as your waistline and other body parts begin to slim down.

Control the eating urge when you feed enzymes to your glands. You can slim down the natural way on the Enzyme-Catalyst Diet Program with wholesome foods and beverages. When enzymes nourish your glands, they can slim down your fat cells and help give you the permanent weight loss dream you are entitled to receive!

Important Facts:

1. Three glands determine your eating levels and can influence your compulsive eating habits.

2. Shirley McB. used a natural "Enzyme-Hormone Tonic" to soothe her thyroid and lose more than 74 pounds.

3. Allan K. tried a 5¢ natural hormone food that corrected his thyroid gland disorder and helped him control his appetite. He also shed some 68 pounds and trimmed down to a neat 34 inch waistline.

4. Rita J. fought and won the "battle of the bulge" and the "chronic eating urge" by using citrus fruits to nourish her adrenal glands and slim her down to a silhouette figure.

5. Burton S. melted 88 stubborn pounds by eating raw seeds and nuts which acted as "time-release, slow-acting enzymes" to soothe his glands and fight fat.

6. Plant foods on an empty stomach act as miracle appetite controllers, while "squeezing" fat out of the body cells and bringing down excess weight.

16

Your Two Week Enzyme-Catalyst Diet "Slim Down" Plan

Set yourself a goal. With the use of enzymes, you will slim down within a two week time span. Each day should be accompanied by healthful physical activity. When your body is active, your enzymes are stimulated to metabolize accumulated carbohydrates, calories and fats. Active metabolism can help create more active enzyme catalyzing of weight-adding substances. You can start your two week ECD "Slim Down" Plan with easy physical activity. Here is how it works.

How To Keep Permanently Slim. To maintain your weight, you need 15 calories for every pound you weigh. If you weigh 150 pounds, you need 2,250 calories a day just to keep your weight where it is. But if you want to reduce your weight, you need to cut 3,500 calories for every pound you want to lose. To lose two pounds a week, you must cut out 7,000 calories a week or 1,000 calories a day. Thus, if you weigh 150 pounds and want to weigh 140, losing two pounds a week, you must reduce your daily intake from 2250 to 1250 for five weeks.

Control Calorie Intake. Your Two Week ECD "Slim Down" Plan calls for controlling your calorie intake. But more important, *you need to improve enzymatic calorie burning through physical activities.* While it is true that enzymes must burn 3500 calories to lose 1 pound of fat (the equivalent of walking to the top of the Washington Monument 49 times), you should remember that walking 20 minutes a day will enable your enzymes to burn up approximately 100 additional calories per day. In a year's time that is 10 pounds. To help your enzymes burn up even more calories, consider steady, moderate exercise, as part of your overall ECD weight-reducing program to help burn up small amounts of extra calories. Over a period of time, these small losses can add up to a loss of extra unwanted pounds. Furthermore, alerted enzymes will also stimulate your circulation, soothe tension, loosen your joints, add tone to your muscles and give you a feeling of well-being. Here is a chart showing you some activities and how they can help melt away calories through activated enzymes:

Exercise-Activity Table

	Calories Per Hour		Calories Per Hour
Bicycling	300-420	Ping Pong	360
Bowling	260	Running	800-1,000
Dancing	450-700	Skating	300-700
Gardening	350	Skiing	600
Golf	210-300	Swimming	350-700
Housework	180-240	Tennis	400-500
Horseback Riding	180-480	Volleyball	210
Ice Skating	360	Walking	100-330

Basic Enzyme Activators. To alert the activity of your enzymes, some basic exercises include walking, gardening, cycling, swimming. Active team sports such as basketball, volleyball or softball also help activate enzymes, especially for younger persons. Calisthenics or weight training may be tedious to some, but enjoyable to those who practice them regularly. The exercise should be sufficiently vigorous to alert your enzymes so they can help use up enough calories to cause weight loss.

Body Weight Controls Enzyme Vigor. Energy expended is also affected by body weight since in those activities where you have to move your own weight, energy costs are increased for the heavier person and decreased for the lighter.

For example, if you weigh 100 pounds and walk 3 mph, your enzymes will burn as few as 50 calories in 15 minutes. Someone else, weighing 200 pounds and walking 3 mph, would use up as many as

80 calories in the same length of time. If you are very heavy, then eat more raw foods so your body will have ample supply of enzymes that can help melt away calories and keep you slim. But the key to enzyme vigor lies in keeping your body active.

The following chart shows how certain activities can help use up excess calories and keep you slim:

ENERGY EXPENDITURE BY A 150 POUND PERSON IN VARIOUS ACTIVITIES*[1]

Activity	Gross energy Cost-Cal per hr.
A. Rest and Light Activity	**50-200**
Lying down or sleeping	80
Sitting	100
Driving an automobile	120
Standing	140
Domestic work	180
B. Moderate Activity	**200-350**
Bicycling (5½ mph)	210
Walking (2½ mph)	210
Gardening	220
Canoeing (2½ mph)	230
Golf	250
Lawn mowing (power mower)	250
Bowling	270
Lawn mowing (hand mower)	270
Fencing	300
Rowboating (2½ mph)	300
Swimming (¼ mph)	300
Walking (3¾ mph)	300
Badminton	350
Horseback riding (trotting)	350
Square dancing	350
Volleyball	350
Roller skating	350
C. Vigorous Activity	**over 350**
Table tennis	360

---

[1]"Exercise And Weight Control," President's Council on Physical Fitness, 1975, Washington, D.C.

Ditch digging (hand shovel) 400
Ice skating (10 mph) 400
Wood chopping or swing 400
Tennis 420
Water skiing 480
Hill climbing (100 ft. per hr.) 490
Skiing (10 mph) 600
Squash and handball 600
Cycling (13 mph) 660
Scull rowing (race) 840
Running (10 mph) 900

HOW TO ALERT ENZYME ACTIVITY THROUGH
SIMPLE CHANGE IN EATING HABITS

Weak or sluggish enzymes may be the cause of improper eating habits. Make a simple change in the way you eat and your enzymes can be self-regulated for stronger fat cell slimming.

How Simple Chewing Helps Lose Weight. Overweight Peggy LaB. was always in a hurry. She had a house to manage as well as a job to fulfill. Eating was something to get over with. She gobbled down her food. Furthermore, she hardly ever did any regular exercise. Other than housework and sedentary office work, she was physically inactive. She started to gain weight until she tipped the scales at a very high 180. Even when she ate less, she was still overweight. The reason? Her digestive enzymes were sluggish because she did not chew her food properly. The act of chewing would help alert more enzymatic creation. Futhermore, with simple physical activity, her enzymes would be invigorated to attack the accumulated food for greater catalyst slimming. Peggy LaB. then followed this two step program:

1. All food should be eaten slower. Chewing should be thorough.

2. After each meal, a 15 minute walk.

Results? Her enzymes were alerted and invigorated and she soon melted some 65 pounds. On this 2-step program, the weight was not only lost, but it was kept off! Enzymes helped keep her fat cells slim, if she chewed thoroughly and then kept physically active.

Simple Corrective Program: Enzymes control your body's satiety mechanism. They let you know when you have had enough to eat.

When you eat rapidly, you eat more food than you normally would before your enzyme satiety mechanism has a chance to react.

To make a simple correction, *eat slowly.* If you are still hungry after you have completed your meal, wait 15 minutes before asking for more. By that time, your enzyme satiety mechanism will have caught up and you should not want additional food. It is the natural way to use enzymes for self-control and reduced food intake.

YOUR TWO WEEK ECD "SLIM DOWN" PLAN

First Day

Breakfast: Fresh citrus fruit juice; poached egg on English muffin; yogurt with fresh fruit slices; coffee substitute or buttermilk.

Lunch: Raw vegetable salad with lemon juice and apple cider vinegar dressing; tuna salad on open slice of whole grain bread; dish of applesauce with sun-dried raisins; beverage.

Dinner: Mixed green salad with dietetic dressing; cooked brown rice; lean broiled steak with lemon or orange slices; sponge cake; beverage.

Second Day

Breakfast: Fresh blueberries in granola with skim milk; half grapefruit; whole wheat toast with fruit puree; coffee substitute or buttermilk.

Lunch: Tomato juice; tuna noodle casserole; Waldorf salad; Fruit whip; beverage.

Dinner: Baked potato; sliced lettuce-cucumber-tomato salad with citrus fruit dressing; meat loaf; oatmeal cookies; beverage.

Third Day

Breakfast: Tangerine juice; broiled finnan haddie; flaky biscuits; assorted fruit and skim milk cheese platter; coffee substitute or buttermilk.

Lunch: Mixed vegetable juice; sandwich with meat filling; lettuce and tomato salad with pinch of oregano, salad oil dressing; fresh fruit wedges; beverage.

Dinner: Mashed potatoes; steamed broccoli; seasonal raw vegetable salad with apple cider vinegar-oil dressing; baked chicken with lemon wedges; baked apple with honey; beverage.

Fourth Day

Breakfast: French toast with orange wedges; fresh fruit slices in yogurt; assorted seeds and nuts; coffee substitute or buttermilk.

Lunch: Gazpacho soup; sliced turkey sandwich with seasonal raw vegetables; pound cake; beverage.

Dinner: Carrot sticks, sliced pickles, celery; liver and onions; steamed string-beans; cantaloupe; beverage.

Fifth Day

Breakfast: Seasonal berries in skim milk cottage cheese; pancakes with honey or maple syrup; fruit slices in yogurt; coffee substitute or buttermilk.

Lunch: Vegetable soup with melba toast; chicken salad sandwich with fresh raw vegetables; tomato juice cocktail; beverage.

Dinner: Broiled halibut steak; baked potato; steamed carrots and peas; lettuce and tomato salad; raisin brown rice pudding; beverage.

Sixth Day

Breakfast: Sliced peaches; scrambled eggs; breakfast muffins; yogurt with sun-dried raisins; dish of cherries and grapes; coffee substitute or buttermilk.

Lunch: Apple-pear-banana salad; grilled cheese-tomato sandwich on protein bread; fruit juice; assorted seeds, nuts; beverage.

Dinner: Raw vegetable salad; London broil with sliced seasonal fruit topping; dinner rolls; sunflower seeds; beverage.

Seventh Day

Breakfast: Muesli (soak 1 tablespoon of raw oatmeal overnight in fruit juice to cover. Next morning, stir in 2 tablespoons of honey and milk, top with apple slices.); cottage cheese with chopped celery, radishes; blueberry muffins; coffee substitute or buttermilk.

Lunch: Fruit compote; cheese-fruit platter; melba toast crackers; beverage.

Dinner: Rice pilaf; sauerkraut and cole slaw salad; roast of veal; pineapple-banana fruit salad; beverage.

Eighth Day

Breakfast: 1/2 grapefruit; whole wheat toast with butter or margarine pat; poached egg; coffee substitute or buttermilk.

Lunch: Assorted sun-dried fruits; tomato juice; tuna salad sandwich on whole grain bread; beverage.

Dinner: Tossed green salad with fruit juice dressing; lean roast beef in natural gravy; steamed asparagus; beverage.

Ninth Day

Breakfast: 1/2 cataloupe; seasonal fruit slices in yogurt; almonds, nuts; coffee substitute or buttermilk.

Lunch: Cup vegetable soup; cheese-lettuce-tomato sandwich on whole grain bread; assorted fruit slices; beverage.

Dinner: Steamed beets; baked salmon with fruit slices; applesauce; assorted seeds, nuts; beverage.

Tenth Day

Breakfast: Glass fig or prune juice; grapefruit; poached egg; whole grain toast; sun-dried raisins; coffee substitute or buttermilk.

Lunch: Cup broth, bouillon or consomme; 4 canned sardines; lettuce-tomato salad with fruit juice dressing beverage.

Dinner: Seasonal raw salad bowl with apple cider vinegar-oil dressing; baked, broiled or stewed breast of chicken; dinner rolls; dish of almonds, nuts; beverage.

Eleventh Day

Breakfast: Oatmeal with fresh fruit slices; yogurt with banana and apple; dish of grapes; coffee substitute or buttermilk.

Lunch: Broiled hamburger patty with onion ring; Waldorf salad with dietetic dressing; seasonal fruit platter; beverage.

Dinner: Fresh raw vegetable salad; baked halibut with fruit slices; vegetable cocktail; assorted seeds; beverage.

Twelfth Day

Breakfast: Breakfast waffles with maple syrup; fruit slices with honey drizzle; almonds, sunflower seeds; coffee substitute or buttermilk.

Lunch: Lettuce-Tomato salad; braised lamb; steamed broccoli; fruit cocktail; beverage.

Dinner: Carrot juice; roast turkey; seasonal vegetable salad with apple cider vinegar-oil dressing; whole wheat bread slices; pineapple slices; beverage.

Thirteenth Day

Breakfast: Natural cereal with sun-dried raisins, pitted cherries, in skim milk; fruit-flavored yogurt; coffee substitute or buttermilk.

Lunch: Cabbage-celery juice; salmon salad sandwich on lettuce leaves with assorted seasonal vegetables; fruit-flavored gelatin with sun-dried fruits; corn muffins; beverage.

Dinner: Steamed squash with lemon juice dressing; Swiss steak with fruit slices; cup mushroom-barley soup; fruit compote; beverage.

Fourteenth Day

Breakfast: Western omelet; cheese and fruit platter; yogurt with wheat germ; walnut meats; coffee substitute or buttermilk.

Lunch: Salad of greens, tomatoes, raisins; lamb chop (fat removed); raw fruit salad with dietetic dressing; beverage.

Dinner: Steamed soy bean platter; raw vegetable salad; dinner rolls; assorted seeds, nuts; beverage.

HOW TO ACTIVATE ENZYMES ON YOUR TWO WEEK ECD DIET "SLIM DOWN" PLAN

Husband and wife, Martin and Angela U. were both overweight. They set upon a program using raw foods as the foundation for boosting their enzyme supply. It was their hope that these enzymes would then help metabolize accumulated weight in the adipose cell tissues and help take off weight and keep it off. They followed the Two Week ECD Diet "Slim Down" Plan, but it was Martin who eventually shed some 38 extra pounds. Angela U. could lose only 3 or 4 pounds. Why did Martin U. lose, but Angela U. still maintain the unwanted extra pounds?

The answer is that after each meal, Martin U. would take a comfortably long walk, keep himself active, so that his body enzymes could be alerted and thereby have the vitality to catalyze accumulated calories in the adipose tissues. But Angela U. would remain inactive, thereby slowing down her digestive enzymes. Once Angela U. learned that metabolism is improved when a short walk is taken after a meal, that this activity is a form of exercise that alerts the catalyzing action of the enzymes, then she was able to shed over 40 excess pounds. Now, both Martin and Angela U. lost weight on the two week program. More important, through the enzyme catalyst response, the weight was kept off, *permanently.*

DO'S AND DON'TS FOR BETTER SLIMMING RESULTS ON YOUR TWO WEEK ENZYME-CATALYST DIET PLAN

To put natural reducing power into your enzyme system, here is a set of do's and don'ts to follow during your two week program . . . and afterwards, too, for better catalyst reaction for slimming down your adipose cells:

1. Eliminate refined sugar and products made with this enzyme-destroyer. Replace with modest amounts of honey, maple syrup, blackstrap molasses.

2. Replace white flour products with whole grain products such as bread, cereals, even macaroni and spaghetti. Whole grains are sources of polyunsaturated oils that help move enzymes with natural lubrication.

3. Select non-hydrogenated fats such as margarines, nut butters. Use polyunsaturated vegetable oils. These are beneficial to your enzymes.

4. Beverages should be *alive* such as raw fruit and vegetable juices. Avoid caffeine-containing coffee and teas. Select substitutes such as Postum, Ovaltine, herbal teas. These soothe the digestive enzyme system so they can function smoothly without being "enslaved" to the caffeine habit.

5. Wherever possible, try to eat wholesome, healthy and natural foods. Preservatives, additives are chemicals which

are destructive to enzymes. Read the label of a packaged product. If it contains a preservative or additive, better pass it up.

6. Continue enjoying lean meats, fish, fowl, but to help in their metabolism, eat fresh raw fruits or vegetables at the same time.

7. Take advantage of the high enzyme content in sun-dried fruits, those that are natural with no sugar or preservatives added. These are high in calories so should be eaten occasionally. They are prime sources of enzymes, however, and should be used as an occasional dessert, especially after a heavier meal.

8. After each meal, try to take a walk to help improve the catalyst action of enzymes. Exercise about one hour after your meal, if you are inclined toward more progressive activity.

9. Weigh yourself daily to note your weight loss on the Two Week ECD Plan. If there is no noticeable loss, then increase intake of *raw* foods and reduce portions of cooked foods.

10. Daily, look at yourself in the mirror. Take note of your shape. As you see yourself growing slimmer every day under the Two Week ECD Plan, you will feel greater motivation to continue eating and reducing!

Let enzymes do your reducing for you! Eat enzyme good foods, keep yourself physically active, and you can enjoy the Enzyme-Catalyst Diet "Slim Down" Plan for more than two weeks . . . for the rest of your slim lifespan!

Summary:

1. To keep permanently slim, activate enzymes with easy exercise and physical actions.

2. Simple chewing helped Peggy LaB. lose some 65 pounds.

3. Enjoy eating on the Two Week ECD "Slim Down" Plan, which gives you good taste and superior enzymes that help slim down your adipose cell tissues.

4. Martin and Angela U. lost weight by energizing enzymes through easy walking after a meal.

5. It's fun to keep slim when you follow the do's and don'ts during the two week ECD Plan.

17

Fifty ECD Slimming Secrets For Daily Weight Loss

Cell-slimming enzymes can work around the clock to help take off weight . . . and keep it off, *permanently*. To put vigor and activity into your digestive system so that enzymes can continue the process of *autolysis* or "loosening" of accumulated carbohydrates, calories and fats from the membranes of the cells and tissues, then helping to dissolve them and "wash" them out of the system, here is a set of 50 little-known but highly-effective secrets for daily weight loss. Many of these Enzyme-Catalyst Diet secrets are so easy, you will hardly know they are being used. Yet, you will discover the scale reading is going down, almost daily, because of these slimming secrets. Use as many of them as possible for better enzyme slimming.

BEGIN WITH THIS BASIC "RECIPE"

The "recipe" for successful ECD slimming is simple:

Take *Motivation*—strong reasons for losing weight

Add *Knowledge*—about the power of enzymes

Mix with *Self-Discipline*—will power

Dash of *Extra Physical Activity*

Put them all together, and they will help slim down your fat cells and give you the healthy, youthful shape you have always wanted.

1. *Enzyme Dressing.* Use fruit juice, lemon juice or apple cider vinegar for an enzyme salad dressing.

2. *Reduce Cooking Fats.* All meat, fish and poultry should be broiled, baked or roasted without additional fats. This makes it easier for enzymes to help break down the fat that will gather in the cells.

3. *Remove Visible Fat.* To make it easier for enzymes to metabolize fats, do not overeat fat. Remove all visible fat from meat before eating.

4. *Go Easy on Sauces, Gravies.* These coat foods and make a thick covering on foods, reducing enzyme digestion. Use these sparingly.

5. *Skim-off Soups.* Pauline T. loved thick, hearty soups. These added heavy body weight. How could she continue indulging in soups, and lose weight? Simple. She skimmed off all visible fat from soups. She also enjoyed clear broth types of soups with additional raw, sliced vegetables for better enzyme metabolism. She lost over 44 extra pounds while she continued indulging in her favorite but "slim soups."

6. *Use Natural Spices.* Avoid salt, pepper, mustard, ketchup, vinegar. These antagonize and destroy enzymes. Replace with flavorful natural herbs and spices such as vegetized salt substitute, paprika, apple cider vinegar and a wide variety available at most supermarkets.

7. *Eat Raw Vegetables.* Eat these daily, with an enzyme dressing of fruit juice with apple cider vinegar, some oil and honey.

8. *Eat Raw Fruits.* Sprinkle with honey, for good taste. Chew thoroughly to stimulate better enzymatic action. Eat daily.

9. *Enjoy Skim Milk Products.* Make it easier for enzymes to slim your cells with less dairy fat intake. Enjoy skim milk products (milk, cheese, yogurt, etc.) available at all supermarkets. All the joyful taste of regular milk but with enzyme-happy lower fat.

10. *Drink Lots Of Water.* Do this *between* meals to help moisturize your cells and make it easier for enzymes to penetrate the adipose tissues for fat metabolism. Do *not* drink heavily with meals since you may "drown" enzymes and inhibit their weight-reducing powers.

11. *Focus Attention On Foods.* When eating alone, devote total attention to your food. Do not read or watch television. This helps boost enzyme-digestive powers and greater cell penetration.

12. *Eat Comfortably When With Other Folks.* When eating with others, do not talk with food in your mouth. This reduces digestive enzyme content. Swallow first. Put down your utensils before talking. Become aware of the difference between eating and talking. Do not mix the two actions. Enzymes can perform when you chew foods thoroughly.

13. *Simple Chewing Keeps You Slim.* Do this very thoroughly so mouth enzymes can prepare food for better assimilation and better metabolism. Do not swallow it half-chewed. Search out the cavity of your mouth with your tongue for large morsels. Reduce these to a smaller size.

14. *Enjoy Food For Enzyme Potency.* Better food enjoyment creates more enzyme potency. Notice the taste, texture and tempera-ture of the food. How sweet it is. How tart it is. How smooth or rough it is. How warm or comfortably cool it is. All of these enjoyments create digestive harmony for better enzyme cellular metabolism.

15. *Sit While Eating.* Nervous Ned I. was overweight because he usually stood or even walked around while munching, nibbling, even eating. He was told that his enzymes were too upset over this unnatural movement and they could not perform their cellular metabolism. Hence, Ned I.'s heavy waistline, jowls, and wheezy breathing because of the overweight. He was told to sit while eating for happier enzyme reaction. He was told to put a small portion of food on a dish, sit at a table, concentrate on eating. This created better enzyme function. This method also helps control over-eating which is part of the syndrome of nervous weight gaining. Ned I. slimmed down to a neat 32 inch waist; his jowls "melted" and his breathing was healthy. He eats satisfactory portions of foods, but does so *sitting down only* for better enzyme metabolism and fat melting benefits.

16. *Eat Less At mealtimes.* Do this by first filling up on any raw fruits, vegetables, seeds, nuts. This gives you a satiety value. Also, you fortify your system with living enzymes for better cellular metabolism. Leave portions on your plate when you've had enough. You can even avoid the temptation to over-eat by *not* loading up your plate in the beginning.

17. *Eat Enzyme Foods Between Meals.* Go ahead and snack. It can keep you slim. But snack on enzyme foods such as carrots, radishes, cucumbers, cherry tomatoes, celery. Also, enjoy their juices. They are prime enzyme sources for cellular slimming. Also helps cut down on over-eating.

18. *Keep Yourself Active.* Alert your sluggish metabolism by activating your enzymes through simple physical energies. For example, walk more and drive your car less. Use stairways, instead of elevators, for one or two flights. Stand up in a subway or bus. Keep your body in motion, your enzymes active, and you'll help keep slim, too.

19. *See Yourself As Permanently Slim.* Make a daily habit of self-conditioning your thoughts so that you visualize yourself as being slim. Emotions can help regulate your enzyme activities. As you look at yourself in a mirror, tell yourself you are getting ten, twenty, thirty pounds slimmer. This positive-thinking attitude can program your enzymes to function smoothly.

20. *Snack Enzyme Foods At TV Time.* Go ahead and indulge in snacking while watching TV. But use enzyme foods. This fortifies your system with these cell-slimming substances while you sleep! A platter of crisp, raw vegetables to snack on, while watching TV before bedtime, can help satisfy your nibbling urge and give you needed enzymes for slimming down.

21. *Instead Of Food, Take A Walk.* Valerie A. liked to raid the refrigerator for second and even third helpings. She kept gaining more and more weight. To control this urge, she decided she would go for a walk, or a casual bicycle ride, whenever she wanted to raid the refrigerator. **Benefit:** She cut down on excess over-eating; the physical activity further helped alert her enzymes for better cellular slimming. She was soon able to slim down to about 130 pounds, on this easy program.

22. *Make A List Of "No-No" Foods.* Whenever you abuse your enzymes with so-called "junk" foods that are a "No-No" on your ECD program, make a list of them. Write down how many cakes, cookies, candies, sweets, etc., you have eaten. At day's end, look the list over. Feel ashamed? Good. It's the natural way to control eating these enzyme destroyers. Next day, look at the list whenever you are tempted. It helps keep weight down and perks up enzyme health, too.

23. *Eat A Hearty Enzyme Breakfast.* Never, but *never* skip breakfast. It is the most important enzyme meal of the day. Eat raw fruits, vegetables, seeds, nuts as part of your breakfast. This fortifies your digestive system with a treasure of enzymes that wait for food to be eaten later in the day and to help burn up unwanted carbohydrates, calories and fats to keep your cells from gaining weight. Always eat breakfast. Always eat some raw fruits for breakfast, too.

24. *Small Plates.* Satisfy your eyes and soothe your enzymes by putting smaller portions on snack-size or smaller plates. This gives you a feeling of ample food and you'll feel satisfied when you finish. So will your enzymes which are grateful for not being overworked.

25. *Buy Slimmer Clothes.* Anticipate losing more weight and being able to wear slimmer clothes. This creates a better attitude. Visualize yourself in the slimmer clothes. Hang them on the inside of your closet door to give you courage to continue on your slimming program.

26. *Keep Yourself In Good Condition.* Keep neat, clean, attractive. Treat yourself to good cologne, perfume, makeup. Both men and women should be good to themselves with new clothes or reward themselves in other ways. Make this a "be nice to me" week or month. Enjoy your favorite hobbies, books, motion pictures. You'll feel contented. Your enzymes will function with more vigor when you are happier and in good condition.

27. *Keep Yourself Busy.* Boredom and monotony can cause nervous tension and inhibit the action of enzymes. So keep busy. It's also the best way to avoid the temptation to eat. Write down a list of all kinds of chores you want to do. Straighten out your closet. Tidy up your shelves or drawers. Do some painting. Try sewing. Do as many pleasurable activities as possible so you won't feel bored. Keep busy. Your enzymes will keep busy, too.

28. *Avoid Eating When Tense.* Nervous tension, fears, apprehensions, arguments, all constrict your digestive mechanism and inhibit the function of enzymes. If you are in the midst of an anxiety situation, then do NOT eat. Instead, relax for about fifteen minutes. When your body is relaxed, your mind will be soothed. Then you should be able to enjoy eating and your enzymes will feel more active, too.

29. *Enjoy A Night's Sleep.* Better cellular metabolism begins with a good night's sleep. Your enzymes keep on digesting foods while you sleep when your body metabolism can devote full concentration on fat melting. It's vital to have a good night's sleep on your slimming program . . . and otherwise, too!

30. *Low-Fat Casseroles Can Be Enjoyed.* Indulge in healthy vegetable, bean, lean meat and fish casseroles. But don't overload your enzymatic system with too much fat. Instead, put a cooked casserole in the refrigerator overnight. Next morning, skim off fat that has risen to the surface. This gives you a tasty low-fat casserole that will not overburden your enzyme system.

31. *Adjust Your Enzyme Rhythm.* Let hunger and satiety work for you by adjusting your enzyme rhythm. It takes 20 minutes from the start of a meal for your stomach to transmit sensations of satiety to your brain. Your enzymes are then working at metabolism. Try to extend your meal for about 30 minutes of eating for longer enzyme activity. During your meal, pause, chat idly, or just rest in between eating different foods. Do this toward he end of your meal, when your appetite is largely eased. Increase the number of pauses and the length of each pause. This helps prolong enzymatic action and offers better satiety, too. It can help keep cells much, much slimmer.

32. *Reserve Eating At Mealtimes Only.* Enzymes work effectively at mealtimes. In-between snacks build calories because enzymes are often recuperating from their mealtime metabolism. So try to avoid eating on occasions other than regular mealtimes. Do not nibble while reading, watching TV, or talking on the telephone. If you *must* nibble, select good enzyme foods such as raw fruits, vegetables, seeds, nuts. You'll be boosting enzyme intake while satisfying your eating urge on a low-calorie selection of these raw foods.

33. *Keeping Slim While Dining Out.* Prepare your enzymes for this occasion. Learn to say "No, thank you," to very heavy and enzyme-fattening foods. Control your appetite by having a cup of bouillon, a dish of gelatin, a raw celery stalk or carrot *before* eating anything else. This helps soothe your appetite and controls the urge to overeat.

34. *Easy Exercises To Boost Enzyme Activities.* Pep up your digestive enzymes with these easy non-exercises throughout your day. Walk to a mailbox some ten minutes away to mail a letter (or pretend to), and walk back home and your enzymes will dispose of 250 calories from your cells. Devote 20 minutes to weed pulling in your garden to enzyme-melt 250 calories. Or dispose of the same amount of calories when you have a 20 minute ping pong game with a friend. You'll boost your enzyme's power with these easy non-exercises.

35. *Reward Your Enzymes.* When your enzymes slough off excess weight, reward them (and yourself, too.) Buy yourself a new luxury item of clothing. Or else, put a dime or a quarter away for each day your enzymes lose weight. When you reach your goal of a new slimness, use this money to get yourself a little gift, a well-deserved vacation, or any trinket that makes you feel happy. Your enzymes will feel good, too.

36. *Be Naturally Sweet To Your Enzymes.* Sweeten your digestive enzymes with a cup of Postum or any coffee substitute (sold at most health stores) or herbal tea, to which you have added a pinch of cinnamon. It eliminates the need for high-calorie sugar.

37. *Add Enzymes To Herbal Teas.* To freshly brewed herbal tea, add a twist of orange or lemon peel. Try a lime twist, also. Good enzyme source added to your tea and helps improve the fat melting vigor of digestive enzymes.

38. *The Safe Fat To Eat.* No need to deprive yourself of fat. It is needed by enzymes which transport Vitamins A, D and E throughout your system while keeping you healthy during weight loss. Try diet margarine which has half the calories of regular butter or margarine. It is also rich in polyunsaturated oils that lubricate your enzymes. Let diet margarine soften to room temperature. It will spread further, give you taste satiety while using less of it.

39. *Buttermilk Is A Fermented Enzyme Food.* Natural ferments in buttermilk stimulate the function of enzymes. Only 90 calories per 8-ounce glass, it is tasty as a beverage. Use as a healthy substitute for sour cream. For a healthy gravy, add it just before serving. Don't boil the gravy or the buttermilk will curdle. Fermented buttermilk helps boost enzyme vigor.

40. *Dairy Food For Fat Substitute.* Buzz creamed cottage cheese in a blender as a substitute for sour cream. Use as a dip for Melba toast or fresh fruit slices. The buzzed cheese is easier to digest and helps promote more digestive satisfaction. Enzymes are healthfully nourished by the protein in this type of cheese used as a substitute for heavier sour cream.

41. *Remove Chicken Skin.* Much of the hard fat is found in the skin of chicken. Remove it before baking or broiling. **Benefit:** the flavor of sauces will be better absorbed into the meat. During digestion, enzymes can better metabolize this type of meat. Otherwise, the skin would interfere and hinder enzymatic metabolism. Great idea for turkey, too.

42. *Enjoy Meat And Keep Slim.* Buy very lean meat. Try round and flank steak. Try very lean ground beef. Cook hamburger well-done to dispose of much of the fat. Always trim away excess fat before cooking and before eating. *Always* eat meat with a fresh fruit dish. Enzymes will then help metabolize the calories and fats of the meat which you can enjoy while keeping slim.

43. *Try Home-Baked Desserts.* Most commercial layer cakes are high in calorie-rich, sugar-containing fillings or frostings. These "devour" enzymes, reducing your supply so cells can grow fat. Instead, try old-fashioned "plain" caked such as gingerbread. Bake it yourself, using no sugar, but a bit of honey. You'll have all the taste of a "rich" dessert that is low in calories but high in nutrition.

44. *Use Enzymes As Thirst Quenchers.* Instead of calorie-high and sugar-containing soda, try enzyme drinks as thirst quenchers. Any fruit juice is good. Try them plain, with an ice cube or two, or mix for various flavor combinations. Fruit juices are prime enzyme sources that activate cellular metabolism while quenching your thirst.

45. *Rock Your Way To Slenderness.* When you rock in a rocking chair, your enzymes use two and a half times as many calories as when you sit. Good for your general circulation, too. Enzymes can burn up one calorie *per minute* when you sit, but 2 1/2 when you rock. So rock your way to a delightful new slenderness.

46. *Try Eating Meals In Courses.* Don't overload your enzymes with too many carbohydrates, calories and fats in one meal. Divide a meal into several courses. For example, you'll find a cup of clear soup, followed by a light main dish, then a final course of a fruit dessert, much more appetite-appeasing and digestive-easing than the same number of calories served in one huge dish.

47. *Think Light, Not Heavy.* Rediscover the superb flavor of fresh seafood, simply prepared with a squeeze of enzyme-rich lemon, lime or grapefruit juice. The seasonal delights of fresh asparagus, the delicious taste of broiled fruit-kissed chicken or turkey, the elegance of fresh fruit for dessert, add up to a "light" meal but with "heavy" satisfaction.

48. *Lunching With Fewer Calories But More Enzymes.* Find it difficult to stay with your diet midday? Try this: take a mid-morning break at eleven o'clock with a cup of fresh fruit juice that is brimming in enzymes. Also, nibble on some fresh fruit. Then "get busy." Take a walk. Shop (not for groceries), clean house, or just keep working on your job—if that's where you are—until say about 2 o'clock. Now, have your lunch. You'll want to keep it light, of course (keep telling yourself) because it's only a few hours until you'll eat again.

49. *Try These Appetite Controllers.* These foods do more than work as appetite controllers. They are high in bulk and water content to help fill you up ... and give you a rich treasure of needed enzymes. Before dinner, try these appetizers: Crisped in water, sprinkled with fruit juice ... carrot curls ... celery strips ... raw turnip sticks ... zucchini fingers ... mushroom slices ... cherry tomatoes ... dill pickles ... green pepper slivers ... raw cauliflowerettes ... shredded cabbage. They set your digestive enzymes churning to help slim down your cells.

50. *For Those Who Have Little Self-Control.* And finally, if you must have sugars and sweets, then cut your usual quantity in half. Keep cutting down until your taste buds are re-educated. Then switch to healthful raw fruits that give you all the joyful good taste of sugar, but with a treasure of enzymes that help keep you slim while you eat.

TAKE A GOOD (AND PRIVATE) LOOK AT YOUR NUDE SELF

That's right. Pose naked every day before a full length mirror. Take a very good look at yourself. What do you see? A flabby, heavy body? Or one that is slowly shedding weight? Look at yourself from all angles. It will help you stick to the Enzyme-Catalyst Diet Program, either from vanity-motivation or disgust-motivation. But it will give you an honest reflection of what you are ... and what you can become once you treat yourself to good food and fat-melting enzymes.

In A Nutshell:

1. Lose weight permanently on the Enzyme-Catalyst Diet Plan with as many of the 50 secrets as possible.
2. Pauline T. eats favorite "heavy" soups without gaining weight. She has lost over 44 extra pounds and has kept them off on ECD Secret #5.
3. Ned I. made a simple adjustment in how he ate and helped get rid of his heavy waistline, jowls and wheezy breathing. ECD Secret #15 worked like a charm.
4. Valerie A. used ECD Secret #21 to slim down to 130 pounds without any effort at all.
5. Daily, take a (private) nude look at yourself to be self-motivated to use enzymes for losing weight.

18

How the ECD Diet Program Solves Those Difficult Weight Problems

Fast, permanent weight loss is possible for those who complain of difficult weight problems. The cell-slimming promise of food enzymes holds t.at this substance can split the molecules which cling to the membranes, then help dissolve them and dispose of them through normal eliminative channels. Even if there has been a history of unsuccessful weight loss under ordinary diets, there can be the fast and (more important) *permanent* reduction by the use of food enzymes.

How Enzymes Correct Cause Of Overweight. So-called "difficult" or "impossible" weight is usually traced to overweight cells. Enzymes cause the *osmotic pressure* reaction wherein the capillaries are penetrated, the accumulated carbohydrates, calories and fats are then metabolized. Enzymes create a permeability into the capillaries, helping to "squeeze" out these weight causing elements through the tiny spaces in the cells, themselves. Enzymes create this osmotic pressure, or a pull which will cleanse the capillaries, getting to the cause of overweight; namely, "fat" cells. When enzymes can reduce the "fat" cells, the reaction is overall body reduction, even if there is a hisotry of so-called "impossible" overweight.

WEIGHT PROBLEM—"BIG BONE" STRUCTURE

ECD Program Solution. Miriam O'H. came from a family with big bone structures. All were tall. All had "large frames" upon which overweight made them look like "stuffed dolls." Miriam O'H. lamented that even if she went on a diet, she still had that "heavy, corpulent" look. Her thighs were plump, her arms were heavy. More embarrassing, Miriam O'H. was developing an unsightly double chin that added years to her appearance. She wanted, desperately, to slim down so she could attend a school reunion and meet classmates of several decades ago. But she wanted to look slim. Her ECD Program called for a high vitamin-mineral program, with emphasis upon minerals. The reason here is that plant enzymes from raw fruits and vegetables would use minerals to keep her bone structure strong, while they would then help slim down the fat in the cells that were closest to her bones. Miriam O'H. followed this simple enzyme catalyst program:

Daily, she would eat large amounts of fresh fruit in skim milk cottage cheese and other dairy products. The enzymes from the fruit would use the minerals from the dairy products to nourish her bone structure, get into the marrow and then promote a catalyst action whereby her "fat-covered bones" would slim down. While her bone structure was kept strong, the fat covering them was slimmed down.

Miriam O'H. lost some 49 overweight pounds this way. Her thighs, arms and "double chin" slimmed down. She had a wide frame, but it had *curves* and it was youthful. The heavy fat was melted away through the fruit-enzyme and dairy-mineral cellular cleansing. When she attended the reunion, everyone remarked upon how youthful she looked and "oh, so slim." A simple and tasty program . . . but it worked fast . . . and permanently.

WEIGHT PROBLEM—NERVOUS FATIGUE

ECD Program Solution. When drastic food reduction takes place, the nervous system reacts. Many a dieter starts to overdrink or oversmoke. Many will chew their nails. Many others develop a "feeling of fatigue" because of reduced food intake. This causes a gnawing urge to resume eating, often the urge calls for overeating and not only is weight returned to the body, but in larger amounts.

To alleviate this problem, the Enzyme-Catalyst Diet Program calls for eating larger amounts of fruits. Enzymes will take up the fructose from the fruits and use it for maintaining good energy. Then, boost the enzyme energy-building powers through regular activity. Keep the body busy. Regular walking, mild exercises, more vigorous around-the-house activity will help increase the circulatory system and this will send more enzyme-carrying fructose to the cells of the nervous system to create energy as well as vitality.

This enables the dieter to feel alert and energetic, even while slimming down. Energy-carrying enzymes from fresh raw fruits can help solve the problem of fatigue while dieting. They also help slim down the cells so overall weight reduction takes place, while you feel alert and active.

WEIGHT PROBLEM–HUNGER HEADACHES

ECD Program Solution. Oscar R. always developed pounding headaches when he went on a severe reducing diet. He was very nervous, jittery, would snap at people upon the slightest provocation. It was a vicious cycle. Either remain overweight and "cheerful," or reduce and become "grouchy" with the pounding headaches. The solution called for a *proteo-zyme* combination. Namely, eat protein foods with raw fruits or vegetables. Oscar R. went on a 30 day program during which he would combine protein (meat, fish, eggs, dairy, peas, beans, nuts) with any *raw* fruits or vegetables. When eaten together, the enzymes from the plant foods would take the protein from the other foods and use it to correct the low blood sugar symptom of "hunger headache." The "proteo-zymes" would regulate the pancreas so it would not secrete excess or a deficiency of insulin. This helped keep Oscar R.'s blood sugar at a normal level. He had no emotional distress symptoms. Also, after 30 days, he slimmed down his waistline to a trim 34. The heavy or "stubborn" weight was gone from his torso. He soon needed new clothes for his permanently slim shape. This simple protein and enzyme combination had slimmed him down while he enjoyed most of his favorite foods.

WEIGHT PROBLEM–FAT RUNS IN THE FAMILY

ECD Program Solution. There are fat families. But fat does *not* run in the family. Rather, the urge to overeat and the habit of hearty eating is often a family heritage that you would like to do without. If you complain you are fat because it is part of the family, then adjust your eating methods. True, you may have inherited more fat cells than slim folks, but you did *not* inherit the weight that is put on these cells. You can slim down your cells and lose weight by using raw enzymes *before* and *after* a meal.

Begin your meal with "filling" carbo-zymes found in raw celery, shredded carrots, cucumber slices, tomatoes, sliced radishes, lettuce. Your main meals should have smaller portions. You'll find that the carbo-zymes in the raw vegetable foods help take the edge off your appetite. But more important, you'll find that they also help metabolize the fat building elements from foods and help control overweight. Many so-called "inherited" chubbies tend to eat between and after meals. This puts on excess weight. To counteract this urge, fortify your system with raw plant foods that offer a satiety value. Also, their enzymes will continue metabolizing carbohydrates, calories and fats so that you have less cellular buildup. With these simple adjustments, using enzyme foods, you can be the "slim member" of the family. Hopefully, others will slim down, too.

WEIGHT PROBLEM–SO-CALLED BODY CRAMPS

ECD Program Solution. Phyllis O'B. always developed painful arm and leg cramps when she lost weight. At times, her leg pains were so distressful, she could hardly walk. So she used this as a plausible excuse for going off her diet. She gained weight and then said she was free from her body cramps. But she wanted to be free from her overweight, too. She turned to enzymes for a solution.

During weight loss, as the fat cells slim down, the skin and muscles constrict and there is a feeling of cramps. Usually, this is temporary. Once the adipose cells are slimmed, the body biological clock readjusts and there is a feeling of flexibility throughout the joints.

While weight is being lost, it is important to use enzymes and fermented milk products. Phyllis O'B. would drink buttermilk, eat yogurt and various cheeses made from skim milk, *together* with fresh fruits or vegetables. The fermented milk products were sources of enzymes that were energized by the plant enzymes to nourish the muscles and nerves in the body to ease so-called pains or constrictions. Enzymes used minerals from the fermented milk products to maintain better water balance and blood stability. This eased the tendency to feel cramps. Phyllis O'B. was able to enjoy fast, permanent weight loss on the basic Enzyme-Catalyst Diet Program, and was free from the recurring body cramps by eating a combination of fermented milk with raw plant foods, available at any local market for a modest cost.

WEIGHT PROBLEM–WATER-LOGGED TISSUES

ECD Program Solution. Sometimes it's not the carbohydrates or calories or fat, but it's water that is trapped in the body tissues that is causing the weight. Known as *edema,* this can be a stubborn weight problem. The problem here is that salt in foods are absorbing too much water and becoming lodged in the skin cells and tissues. Salt acts as a sponge. It absorbs water. Then it attaches itself to the adipose cell tissues and water buildup causes overweight. This can be a health risk, too.

Enzymes can be used to help alert the kidneys to metabolize the accumulated salt. Begin by selecting salt-free foods. Packaged foods almost always contain excessive amounts of salt. Read labels. If the product has salt, pass it up. Select *natural* and non-processed foods, available everywhere, such as fresh fruits, vegetables, seeds, nuts, grains. Increase intake of fruit since these enzymes are especially potent and can help free the "locked-in" salt and stimulate the kidneys to help wash out the liquid from the cells. A one day fruit fast, during which fruit enzymes can work without interference from other foods, can be helpful in promoting fast, water-weight loss.

WEIGHT PROBLEM—"QUIT SMOKING-EAT MORE"

ECD Program Solution. Herbert E. was a chain smoker. The heavier he smoked, the lighter he weighed! When he gave up smoking, he started to gain weight. It was a paradox. He called it a vicious cycle. Which was the lesser of the two health evils? Over-smoking, or over-eating? Herbert E. would say that whenever he quit smoking, he would eat more. He felt trapped. He needed to lose weight fast, but he also wanted to free himself from the smoking habit.

The secret here is to use enzymes to adjust the metabolism. Actually, *when you stop smoking, your metabolic rate is reduced.* This can cause weight gain even if you ate no more than you did when you were smoking. The trick here is to *increase* metabolism so that carbohydrates, calories and fats do not accumulate in your cells.

Herbert E. would boost his raw food intake, from the moment he stopped smoking. Enzymes alerted his metabolism, kept it working at a healthy rate. He continued eating most of his favorite foods, even after he stopped smoking, but there was no weight gain. Enzymes promoted a cellular metabolism that "burned up" the weight in his cells and he was permanently slim . . . and free from the smoking habit, too!

WEIGHT PROBLEM—THE URGE TO SNACK

ECD Program Solution. Go ahead and snack . . . but use healthful, tasty snacks! They should have a double-barrelled benefit. Namely, snacks should satisfy your eating urge. Snacks should then give your body needed enzymes for fat-melting in the cells. When you want to snack, you can indulge in these weight-melting goodies:

- Juicy oranges, crisp apples, strawberries, watermelon
- A Fruit kabob. Put a few pieces of fruit on a popsickle stick.
- Peanut butter, oatmeal or raisin cookies with milk or fresh fruit juice.
- Combine peanuts, raisins, wheat cereal served in small paper cups.
- Raw carrot curls, celery sticks, green pepper strips, tomato wedges served attractively on a tray.
- Combine skim milk with peanut butter or a mashed banana; shake vigorously in a jar with a tight cover. An excellent high enzyme-protein-mineral-vitamin bedtime snack.
- Chunks of bananas on an ice cream stick, rolled in wheat germ.
- Washed celery stalks, nut butter and softened cottage cheese.
- Chunks of fresh in-season fruit like apples and pears, skewered on wooden sticks, dipped in orange and pineapple juice and sprinkled with sesame seeds.
- Washed raw green beans, filled with cottage cheese, cabbage slices, cauliflower buds, green pepper strips and carrot sticks.

These are snacks that are prime sources of enzymes that alert your circulation and improve your metabolism. Also, while they help keep your cells slim, they satisfy your eating urge *without* adding too many carbo-caloric fats. It's the tasty way to keep slim.

WEIGHT PROBLEM—BASIC URGE TO OVEREAT

ECD Program Solution. You have this compulsive urge to overeat because of a weak enzyme supply in your system. Simply speaking, there are certain cells in the middle part of the brain which are extremely sensitive to very small changes in the amounts of glucose and certain hormones in the blood as it travels through the hypothalamus. These cells are known as *glucoreceptors* and their function is to control the amount of food you eat. Enzymes are needed to nourish these *glucoreceptors* to create a self-regulatory mechanism that can help control the eating urge. Enzymes will regulate the amount of glucose or hormone used by the hypothalamus. Once the blood levels of glucose are balanced, through the activity of enzymes, there is a natural stopgap on the appetite and the eating urge is naturally abated. Fresh plant foods are prime sources of these enzymes needed by your *glucoreceptors* to maintain glucose-hormone balance and natural appetite control. Fruit enzymes help solve this "overeating" urge.

WEIGHT PROBLEM–REDUCING WHILE OTHERS EAT HEAVILY

ECD Program Solution. When you are dieting, the sight and smell of food can be very tempting. This is especially noticeable in a family where you are the lone dieter while others eat heartily of foods you want to avoid. The sight and smell can also alert your enzymatic system in anticipation of the pleasures of eating. This causes oral and digestive churning that builds up into an obsessive urge to break your diet and obey the impulse! To help soothe your enzymes, give them something else to "chew" upon. When you are reducing, and others are eating heavily, feast upon these slimming enzyme foods:

- Crisp raw vegetables: cauliflower, carrots, celery sticks, raw turnip slices served with a sprinkle of garlic powder and vinegar-oil dressing makes a high-enzyme, low-calorie gourmet appetizer . . . and satisfies your digestion so you will eat less dinner.

- Dry skimmed milk can substitute for whole milk in any cooked dish. It is a powerhouse of vitamins and minerals needed by enzymes to maintain cellular metabolism and balance.

- Lemon juice, rather than butter, is the gourmet's choice for any broiled vegetables. High in enzymes and taste and low in calories.

- Potatoes, mashed with skim milk cottage cheese plus a bit of parmesan cheese will give you good enzyme digestion, rather than the heavier mashed-with-butter-and-cream.

- A cup of clear bouillon with a side dish of chopped, raw vegetables helps soothe your digestive enzymes and appetite, while giving you enzymes and few calories.

- Fish, chicken, turkey or broiled liver can be as satisfying and as nutritious as "red meat." But pound for pound, the calorie count is far less. These meat foods are high in polyunsaturated fatty acids which are needed by enzymes for better cellular metabolism.

- A vegetable souffle—cabbage, carrots, celery—is deceptively filling. Just as satisfying as potatoes or rice and much lower in calories. Good source of vitamins and minerals needed by enzymes for activation within the metabolic system.

These few adjustments give you high-enzyme foods that help keep you slim, while you can eat with pleasure, amongst others who eat their fattening foods. The big difference is that your enzyme foods are tasty and slimming while their heavy foods are overburdening and fattening!

WEIGHT PROBLEM—SKIN SAGS
AS WEIGHT IS LOST

ECD Program Solution. A common problem with weight loss is that there is a noticeable sagging of the skin. Sagging shows wrinkles, furrows, aging or crepe-like folds in many parts of the body. Many folks try to reduce, only to discover their skin is sagging in a way that is so unsightly, they would prefer to regain weight and look "filled out." Enzymes can help protect against this condition. The solution here is to keep your skin cells nourished with good protein foods such as lean meat, fish, eggs, cheese, peas, beans, nuts.

Eat a protein food with a raw fruit or vegetable or a fresh juice. **Benefit:** The plant enzymes will take the protein from the food, transform it into usable amino acids. The enzymes will then use the protein to build the membranes of the *mitochondria,* the nucleus or power center found in the *cytoplasm,* the body of the cell. Enzymes rebuild the walls of the *mitochondria,* firm up the membrane, replenish the destroyed tissue, promote the manufacture of collagen, the glue-like substance that supports the structures of the cells. A combination of *protein* with a *plant enzyme* can help create this structural rebuilding to help guard against breakdown and the problem of "orange peel" skin.

Most reducing programs consist largely of cooked foods. A deficiency of raw foods means a deficiency of enzymes. The protein cannot be activated into rebuilding the cellular membranes and breakdown, destruction can cause the skin to become furrowed and wrinkled. The Enzyme-Catalyst Diet Program calls for fresh, raw foods and their juices to provide enough catalysts to transform protein into amino acids for skin rebuilding. A deficiency of raw foods means that eaten protein cannot be fully metabolized. Starved tissues create the wrinkles that makes reducing a regret instead of a joy. Feed your tissues with enzymes and you will look and feel smooth and youthfully slim . . . all over.

You can shed those "impossible" pounds not only fast, but permanently when you fortify your body with Nature's powerful cell slimmers—enzymes. Available in raw foods everywhere, they are your key to a forever slim shape . . . no matter how difficult your weight problem may appear to be.

Highlights:

1. Enzymes can solve overweight problems. Use them for so-called "impossible" weight gaining situations.

2. Miriam O'H. used enzymes to lose 49 overweight pounds, even if she had a "big bone" structure.

3. Oscar R. prevented hunger headaches while dieting with the use of enzymes.

4. Phyllis O'B. slimmed down, without any so-called body or leg cramps, thanks to enzymes in everyday foods.

5. Herbert E. quit smoking, ate more . . . but lost weight under the simple program of boosting raw food intake and enzyme metabolic adjustment.

6. With the use of raw enzymes in everyday foods, even the most stubborn of overweight problems can be solved.

19

Fifteen Different ECD Programs for Fast, Permanent Weight Loss

Enzymes are Nature's catalysts that work within your metabolic system to reduce your weight fast and to keep it off, *permanently*. When you fortify your system with these catalysts, you can help reduce the accumulated weight-building carbohydrates, calories and fats from your cells, and thereby bring down your overall weight. It is important to use the Enzyme-Catalyst Diet Program to slim down to help improve your health. Every extra pound of weight is a threat to your heart, your arteries, your blood pressure, your organs and your very life. When enzymes slim your cells down, you can enjoy a more vigorous, more healthy and more attractively longer life.

Wide Variety Of ECD Programs. Enjoy a wide variety of different diets throughout your slimming program. This will give you taste thrills galore, while raw food enzymes help slim down your cells. You may find it helpful to change from one ECD Program to another after a few days or a week, for greater taste variety. The choice is up to you. Each of these 15 programs are set up to boost cellular metabolism through enzyme foods, to help you acquire a slim-trim shape and a new outlook in youthful, healthful living.

#1—LOW-FAT, HIGH-ENZYME DIET

Enjoy These Foods: Meats: Broiled, roasted, boiled or baked beef, lamb, veal, beef liver, chicken, turkey. Remove all visible fats. *Seafood:* Cod, flounder, haddock, whitefish, white perch, carp. *Vegetables:* Fresh squash, cabbage, broccoli, cucumbers, green peppers, radishes, turnips; steamed, baked or boiled potatoes, green beans, asparagus, eggplant. *Fruits:* All fresh fruits and fruit juices. *Dairy:* Fermented milk products such as buttermilk, yogurt, cottage cheese, farmer cheese, pot cheese. Skim milk. A maximum of three eggs weekly. *Grains:* Whole grain breads, rolls, brown rice, whole grain macaroni, noodles. *Miscellaneous:* Honey, gelatin, sweet pickles, coffee substitutes such as Postum, herb teas, sun-dried fruits. Use desserts made without sugar or cornstarch.

Restrict These Foods: All fatty and processed meats such as bologna, frankfurters, pork, salami, sausages, smoked meats, duck, goose. No smoked or fried foods of any kind. No French-fried potatoes or processed potato chips. Restrict butter. (Use polyunsaturated margarine.) Avoid olives, commercial mayonnaise or salad dressings, cakes, cookies, doughnuts, pies, pastries, chocolate or cocoa.

ECD Benefit: Without the interference of excessive amounts of fats, enzymes from the plant foods are able to enter into the membranes of the adipose tissues and help dissolve accumulated weight-causing substances. Fat is often an impediment to enzyme cellular metabolism and hinders the catalyst action. A low-fat, high-enzyme diet that is followed a few days per week, can do much to adjust the metabolic system and promote better weight loss.

#2—LOW-CHOLESTEROL, HIGH-ENZYME DIET

Enjoy These Foods: Meats: Fat-trimmed, lean cuts of roasted beef, lamb, chicken, turkey. Enjoy broiled or baked meats, too. *Seafood:* Flounder, haddock, codfish. *Vegetables:* No limits to any vegetables you desire. Raw vegetables are a "must" on the cell-washing program. *Fruits:* All fresh fruits and fruit juices. *Dairy:* Skim milk, cottage cheese, farmer cheese. Margarine made from polyunsaturated oils. *Grains:* Whole grain breads, cereals, brown rice, whole grain macaroni, noodles. *Miscellaneous:* Vegetable oils in salads, egg whites in cooking, honey, coffee substitutes such as Postum, herb teas, sun-dried fruits. Use desserts made without sugar or cornstarch.

Restrict These Foods: Animal fats in any form. Restrict "organ meat" which is high in cholesterol. Avoid pork products of any sort, processed meats. No shellfish. No butter, egg yolk, dairy products made from whole milk, sweet or sour cream. Avoid any fried foods. Avoid chocolate, cocoa, pastries, cakes, pies, etc.

ECD Benefit: Cholesterol is a fat-like substance that clings to the walls of the cells and tissues. It is involved in problems of coronary arterial disorders. Cholesterol "coats" the adipose cell tissues and "seals in" accumulated carbohydrates, calories, fats. This builds up weight. On this Low-Cholesterol, High-Enzyme Diet, the reduction of hard fats enables the enzymes to penetrate the adipose cell tissues and perform needed metabolism and slimming. Unlike fats, cholesterol often becomes clogged and "thick" and requires super-enzyme metabolism for loosening of the unhealthy particles. Enzymes can work better in cholesterol metabolism if you limit intake of animal foods. This causes reduction of excessive cholesterol and general weight reduction. Follow this diet for several weeks to help bring down stubborn weight. To keep it off, restrict intake of excessive animal fats.

#3–CONTROLLED ENZYME "FASTING" DIET

Enjoy These Foods: Any fresh raw fruits or vegetables. Eat these as often as you want in any amounts that you want. Follow this diet for several days. Chew very vigorously to stimulate enzymes.

Restrict These Foods: Any other foods of any sort.

ECD Benefit: Controlled fasting frees your enzyme metabolic system from interference of any other foods, other than raw fruits or vegetables which are prime sources of catalysts. Your entire body clock system becomes readjusted and better precision-timed to perform needed cellular metabolism. This process creates *autolysis* or self-loosening of accumulated weighty substances that are clinging to your cells and tissues. Uninhibited, the enzymes can work totally upon the objective of slimming down your cells . . . and slimming you down, too.

#4–RAW JUICE ENZYME "FASTING" DIET

Enjoy These Foods: Fresh raw fruit and vegetable juices. Throughout your day, drink these as often as you desire. You might have a raw juice assortment for breakfast. For luncheon, have several mixed vegetable cocktails. For dinner, several different vegetable juices. For a nightcap, a favorite vegetable juice.

Restrict These Foods: Any solid foods of any sort. Any cooked foods.

ECD Benefit: Enzymes in fresh raw juices are pre-digested, and within 15 to 30 minutes after being consumed, go to work in helping to dissolve accumulated weighty substances from the cellular network. Raw juice enzymes are speedily assimilated and quickly alerted to create the weight-melting catalyst action required for permanent weight loss. Roberta V. had "stubborn" weight that could not be dislodged. Her problem was a sluggish metabolism. She needed to activate her circulation so that enzymes would become invigorated to perform their catalyst action. She went on the tasty Raw Juice Enzyme Fasting Diet for several days. Without the blockage of fat metabolism, the enzymes could work freely and totally upon the sole task of cleansing the fat cells. This adjusted her metabolism. Roberta V.'s circulation was now activated with the intake of enzymes from raw juices. This sent strong enzymes to her cells and her weight reduction was a success. Now, once or twice a week, she enjoys a Raw Juice Enzyme Fasting Diet for permanent weight loss.

#5—LOW-CALORIE, HIGH-ENZYME DIET

Enjoy These Foods: Smaller portions of lean meats, soft boiled eggs, three times weekly, skim milk and skim milk cheeses. Daily, have large raw vegetable salads. Be moderate in fruit intake. Salad dressings should consist of apple cider vinegar with a bit of oil and desired herbs. Throughout the day, drink coffee substitutes, herbal teas, with a bit of honey.

Restrict These Foods: Whole grain products, cakes, pastries, confections. Avoid all refined foods.

ECD Benefit: When there is a lowering of calories, enzymes can take over the energy-creating process and thereby promote better metabolism of the cells and tissues. Often, calories and enzymes "compete." Calories can add weight while enzymes reduce it. By lowering calories and increasing enzymes, this ratio will help promote more effective, vigorous weight loss, especially for stubborn situations.

#6—CARBO-ZYME FASTING DIET

Enjoy These Foods: Fresh, raw vegetables in any assortment, throughout the day. For a taste variety, enjoy some steamed vegetables. The emphasis should be on fresh, raw vegetables with a tasty salad dressing of fresh fruit juice with polyunsaturated oil. Make raw vegetables your three or more daily meals.

Restrict These Foods: Any foods that are not vegetables!

ECD Benefit: George R. B. may have lost weight, but he could not keep it off. Furthermore, he complained that most diets gave him a "hollow" feeling as well as constant fatigue. His problem was a loss of energy because of a restriction of plant carbohydrates. On this Carbo-Zyme Fasting Diet for two days, George R. B. was given natural energy as enzymes used plant carbohydrates to alert his body systems. Also, enzymes were, themselves, activated by the energy-creating plant carbohydrates and "attacked" accumulated weight in the cells. George R. B. found that his waistline went down, his double chin became "single" again, his thighs slimmed down, when he followed a 1 or 2 day Carbo-Zyme Fasting Diet. Afterwards, he felt great. He was alert and active and brimming with energy. He had less of a temptation to overeat. It helped him keep weight off!

#7—8-MEALS-PER-DAY ENZYME "NIBBLER'S' DIET

Enjoy These Foods: Fresh raw fruits and vegetables, assorted seeds, nuts, uncooked grains such as wheat, sesame seeds, sprouts, smaller portions of lean meats, coffee substitutes, herbal teas, raw juices, skim milk dairy products. HOW TO FOLLOW DIET: Divide your meals into 8 separate sessions. Then, eat smaller portions of your foods during these 8 sessions.

Restrict These Foods: Any refined or processed foods of any sort.

ECD Benefit: On the nibbling diet, enzymes appear to work more vigorously than if subjected to three heavy meals. Small, frequent meals enable enzymes to deplete accumulated fat in the cells. Enzymes then use the metabolized free fatty particles for speedy energy and a feeling of satiety that controls the urge to overeat. Enzymes seem to help "shrink" the stomach and ease pressure on the circulatory systems. On an 8-meal nibbler's diet, enzymes help strengthen your lean muscle mass, and lower your fat deposits. You can take weight off *permanently* when you restore better metabolism by dividing meals into 8 *smaller* portions. Do not fool your digestive system. Do NOT eat 8 heavy meals. This is self-defeating. Rather, eat 8 *smaller* meals for better enzymatic cellular action.

#8—LOW-SALT DIET

Enjoy These Foods: Fresh fruits, vegetables, lean meats, poultry, eggs, whole grain breads (if label says it is low-salt or salt-free) dairy (if low-salt or salt-free), cereals (low-salt or salt-free) and any product which says "salt-free" or "no salt added."

Restrict These Foods: Salt from the shaker. Any processed, packaged food which has salt added. Avoid any foods which contain sodium chloride or any chemical with the word "sodium" which indicates it is salted.

ECD Benefit: When you eliminate refined salt from your food program, you help open up your arteries and permit enzymes to receive oxygen nourishment. These oxygenated enzymes can then be propelled to the capillaries where they can help catalyze weight from the cells. Further benefits include a better mineral balance so enzymes can work freely to help reduce the fat in your cells and help keep weight off... permanently. (Salt consumption narrows the passageway of the small arteries, blocking the free movement of enzymes. Salt also overworks the hormone-producing glands so less energy is available for catalyst action. Salt also causes swelling-congestion in the walls of the arterioles (tiny arterial vessels), inhibiting the free movement of oxygen-carrying blood. This "chokes" off enzymes. Cells can become overburdened with accumulated carbohydrates, calories and fats. Salt also acts as a "blockage" barrier to restrict enzyme metabolism in the cells. So eliminate salt and your enzymes can work freely to keep your cells slim—and you, too!)

#9—FERMENTED MILK DIET

Enjoy These Foods: Buttermilk, yogurt, home-made sauerkraut. You may also have coffee substitutes such as Postum and herbal teas; also enjoy plain bouillon and consomme. Devote one day to this simple diet.

Restricct These Foods: All other foods, except those listed above.

ECD Benefit: Enzymes in the fermented milk products will restore the natural bacteria of the digestive-intestinal tracts and give them a refreshing energy so they can help metabolize accumulated weighty substances in the cells. Specifically, fermented enzymes from these products can be propelled toward the molecules aand then between the walls of the membranes and dissolve carbohydrates, calories fats and washes them out through the eliminative systems. A one or two day Fermented Milk Diet during which nothing else is taken, can help catalyze stubborn weight and promote desired reducing.

#10—CITRUS FRUIT AND CHEESE DIET

Enjoy These Foods: Cottage, farmer or pot cheese made from skim milk. You may also have coffee substitutes or herbal teas with a bit of honey. Also have any citrus fruit in season or combination.

Restrict These Foods: All other foods, except those listed above.

ECD Benefit: Enzymes in oranges, grapefruits, tangerines, lemon-lime wedges will take up the minerals and protein from the dairy foods and use them to wash away sludge from the cells. Enzymes will then use these dairy minerals and protein (of a fermented variety) for rebuilding the cells and membranes. Enzymes will also take the citrus fruit's Vitamin C to create *collagen* substance that helps keep the molecules strong and firm so there is no breakdown during weight loss. A one or two day Citrus Fruit And Cheese Diet can help free those stubborn pounds accumulated in the cells and help create fast and permanent weight loss.

#11—MEAT AND FRUIT DIET

Enjoy These Foods: Lean meat such as beef, veal, chicken, turkey. Trim away all fat before baking, broiling or roasting. Trim away all fat before eating. Any fresh raw seasonal fruit, either singly or in a combination platte1

Restrict These Foods: All other foods, cexept those listed above.

ECD Beneift: Raw fruit enzymes will take the *complete* protein and use *all* of the amino acids for rebuilding your molecules. Enzymes are activated by the complete amino acid pattern in the meats to help "scrub" away debris and then rebuild the cells. You can have all the eating pleasure of juicy good meat with less weight buildup *provided* you eat fresh raw fruits *with* the meat so the enzymes can metabolize the calories and fats. This simple diet plan helped Adele LeC. shed some 44 pounds, with nary a wrinkly or sag on her skin. She loved the delicious taste of freshly prepared broiled meats. But it kept adding so many pounds she did not know how to continue enjoying the pleasures of meats while keeping slim. All she did was devote one or two days to eating meats *with* fruits to satisfy her appetite for the rest of the week. She then had little desire for meats and was able to eat modestly, with satisfaction, as the pounds were melted. Most important, she could enjoy favorite meats with fruits while keeping slim.

#12—FISH AND ENZYME FOOD DIET

Enjoy These Foods: Fish, either broiled or baked. Any raw vegetable or fruit.

Restrict These Foods: Any shellfish. All other foods, except those listed above.

ECD Benefit: Plant enzymes will take the polyunsaturated oils and essential fatty acids from seafood and use them to "wash" the "dry" cells to which wastes and weighty substances have accumulated. Molecules become healthfully moisturized by this process and there is a freer exchange of nutrients and metabolism to help control weight.

#13—RAW FRUIT ENZYME DIET

Enjoy These Foods: Fresh, raw fruits, either singly or in any desired combination, throughout the day.

Restrict These Foods: All other foods, except those listed above.

ECD Benefit: A one or two day controlled fasting program solely devoted to fresh raw fruit will help give your metabolism a treasure of uninterrupted enzymes. *Chewing* the raw fruit sets off an enzymatic chain reaction that proceeds downward into your digestive system. Here, more enzymes are called forth to help metabolize accumulated weights from the cells and tissues. When your digestive system is treated to raw fruit enzymes exclusively, *without* being diverted by other foods, they have a thousand fold power of dissolving carbohydrates, calories and fats from the tissues and helping to create miracle and speedy slimming. Devote one or two days per week to the Raw Fruit Enzyme Diet for fast, permanent weight loss. Delicious, too!

#14—BANANAS AND FERMENTED MILK DIET

Enjoy These Foods: Bananas and any fermented milk product such as buttermilk, yogurt and skim milk cheeses.

Restrict These Foods: All other foods, except those listed above.

ECD Benefit: Banana enzymes are reported to be from 96 to 99.5 per cent utilizable during the digestive process. The banana enzymes will take the fermented enzymes from the dairy product and use them to help wash away accumulations in the molecule. Banana enzymes also spare protein; this means, your skin cells and tissues remain firm and free from the threat of collapse which could produce wrinkles and aging furrows. Enzymes will use the protein to help maintain a cellular and pectin meshwork that creates molecular slimming and strong cellular structure, too. The fermented milk enzymes will be used by the banana enzymes for more catalyst vigor and energy. Banana enzymes will also provide a pectin and delicate fiber network that adds to gastrointestinal bulk. This is the helpful body balance needed, to promote better catalyst action. One or two days on the Bananas And Fermented Milk Diet can do much to provide nourishment as well as cell slimming and cleansing reactions.

#15—FAST, PERMANENT WEIGHT
LOSS ALTERNATING DIETS

Enjoy These Foods: Select any of the preceding 15 ECD diets and follow permitted foods as listed.

Restrict These Foods: All of those foods listed as restricted in the selected ECD diet.

ECD Benefit: Select 7 different ECD diets. One week is devoted to one diet. The second week is devoted to the next diet. So on for 7 weeks. Or, select one ECD diet for each day. Alternate any of the preceding 14 ECD diets. This gives you a *variety* so that you can enjoy good foods while you slim down. Continue rotating or alternating these diets, according to your taste. This helps give you different types of enzyme reactions so that your cells and tissues are constantly being treated to new and varied methods of cleansing. It helps give you day-after-day taste thrills with so many healthful and delicious foods that take your weight off . . . and keep it off . .,. while you eat to your satisfied contentment.

ENZYMES HOLD KEY TO
"FOREVER SLIM" SHAPE

Raw food enzymes are the keys to taking off weight and keeping it off so you can have that "forever slim" shape you have always wanted. These raw food enzymes will penetrate thick layers surrounding your adipose cell tissues and then dissolve the substances that are causing weight buildup.

Build enzymes into your daily food plan and you can achieve fast, permanent weight loss.

Important Points:

1. Set your weight goal. Plan your daily eat-and-reduce through enzyme catalyst action with a variety of 15 different diet plans.

2. Roberta V. lost "stubborn" weight on a tasty raw juice enzyme program.

3. George R. B. ended dieting problems of feeling "hollow" or fatigued on a delightful Carbo-Zyme Fasting Diet. His waistline trimmed down, his double chin became "single" again,

his thighs were slim, too. He was a bundle of energy on the slimming enzyme diet plan.

4. Adele LeC. shed some 44 pounds, yet had a firm wrinkle-free skin, on a meat and fruit diet program.

5. Enjoy fast, permanent weight loss by alternating these 15 diet plans to suit your taste and your waistline, too!

20

Your Enzyme-Catalyst Diet Cookbook for Keeping "Forever Slim"

You can indulge in a variety of tasty foods while you slim down to a youthful shape. Basically, you need to remember one rule: *use a fresh raw food with each and every meal.* You may use fresh raw fruits, vegetables, raw juices, seeds, nuts, uncooked grain products such as wheat germ as well as fermented foods such as yogurt, buttermilk. These raw foods will give your adipose cell tissues the enzymes they need to help catalyze away the accumulated weight substances from your body. These raw food enzymes will help keep you "forever slim" while you continue eating delicious foods.

Your body has its own "build in" reducing station. Namely, your enzymatic system. Feed it healthful foods. Alert it with food enzymes. Your cells will become *permanently slim* . . . and *so will you!*

RECIPE SECTION

High-Enzyme, Cholesterol-Free Mayonnaise

Mash banana in a small bowl. Beat in polyunsaturated oil gradually until thickened. Add a little lemon juice. Add chopped red cherries. Blend together. Use as a delicious, high-enzyme, cholesterol-free mayonnaise.

Fermented Salad Dressing

> 1 cup yogurt or buttermilk
> 2 tablespoons lemon or any citrus fruit juice
> 1 teaspoon diced onion
> 1/4 teaspoon honey

Blend together all ingredients. Then spoon over any raw salad as a healthy enzyme-high dressing.

Cottage Cheese Dip

> 1/2 cup skim milk cottage cheese
> 2 tablespoons citrus fruit juice
> 1/4 teaspoon honey

Combine all ingredients. Beat until smooth. Add a little dill or paprika for color. Makes a healthy dip.

Fruit Salad Dressing

Combine equal amounts of honey and citrus fruit juice for a fruit salad dressing.

Golden Dressing For Fruit Salad Platter

> 1/2 cup pineapple or orange juice
> 1/4 cup lemon or lime juice
> 1/4 cup honey
> 2 egg yolks

Put all in a small saucepan. Steam very lightly over low heat until thickened. Offers a delicious and golden dressing.

Proteo-Zyme Dressing

> 1/4 cup natural peanut butter
> 1/2 cup skim milk
> 1 tablespoon honey
> 1 tablespoon citrus fruit juice

Blend all together thoroughly. Serve over fruit salads. High in protein and enzymes.

Enzyme-Booster French Dressing

> 1/2 teaspoon honey
> 3/4 cup polyunsaturated vegetable oil
> 1/2 teaspoon paprika
> 2 tablespoons citrus fruit juice
> 2 tablespoons lemon or lime juice
> 1 teaspoon grated rind

Shake all ingredients together. Use as a healthful enzyme-booster French dressing.

Cabbage Salad Dressing

1/2 cup buttermilk or yogurt
1/4 cup honey
1/4 cup lemon or lime juice
1 teaspoon celery, dill or caraway seed (or mixture)

Mix well. Then add 1/2 cup diced cabbage. Use as a salad dressing.

GOLDEN DOZEN ENZYME SALAD RECIPES

1. Create enzyme salad bowl from eggplant shells, avocado halves, pineapple shells, cucumber canoes, large red apple scooped out, cantaloupe halves, watermelon shells, a large grapefruit or orange shell, red or green pepper shells.

2. Use grated or chopped radish in place of hot peppers.

3. Grated nuts or a sprinkling of seeds make an excellent enzyme topping for raw fruit and vegetable salads.

4. Use chopped raisins and nuts on top of salads. Otherwise, when used with a salad mixture, they may darken the ingredients.

5. Garnish the edge of lettuce, apple, pear, pineapple slices by sprinkling with paprika.

6. Flute cucumbers and bananas by running a fork the length of them all around before slicing.

7. Sprinkle your salad with fresh raw spices and herbs for healthful taste and enzyme boosting powers.

8. Pineapple wilts fresh strawberries. Do not mix these fruits. Arrange them in separate mounds.

9. Try not to crush leaves since this may cause enzyme evaporation. To conserve enzymes and other nutrients, tear or cut leaves with a sharp knife or scissors.

10. Use a little wheat germ in cabbage, cucumber and carrot salad to absorb the excess juice which is rich in enzymes.

11. Use a variety of different salad greens: seasonal crisp lettuce slices, watercress, cabbage, dandelion, nasturtium leaves, small celery leaves, young beet tops, chives, parsley, kale, mint, with a sprinkle of sage, dill and basil.

12. Boost enzyme vigor in salads with an accompaniment of seeds, nuts, almonds.

GRATED CARROT

2 cups grated carrots
1 cup chopped celery
1 teaspoon fruit juice
1 teaspoon polyunsaturated oil

Mix together and use as a salad with a main dish.

Celery-Nut Loaf

2 cups ground or grated celery
1 cup peanuts or almonds, finely ground
1 mashed avocado
2 tablespoons diced onion
2 tablespoons minced parsley
1/2 teaspoon thyme
Juice of one small lemon or lime
2 tablespoons home-made mayonnaise

Mix all of the ingredients very thoroughly and place in a bowl.

Sauce:

4 washed, sliced tomatoes
Juice of one small lemon or lime
1 teaspoon honey
1/2 teaspoon marjoram and thyme

Blend together all of these ingredients. Use as a high-enzyme sauce topping for the Celery-Nut Loaf which is high in enzymes and protein, too.

Cookless Potato Salad

1 medium Irish potato, skin-scrubbed, then shredded
1 teaspoon parsley
1 teaspoon polyunsaturated vegetable oil
1 teaspoon chopped onion
1/2 mashed avocado

Blend all ingredients together. Sprinkle with some paprika or herbs. Serve as a high-enzyme cookless potato salad.

Protein-Enzyme Loaf

1 cup scrubbed carrots
1 cup tomato wedges
1/2 cup washed parsley
1 cup green pepper, minced
1 clove garlic
2 tablespoons vegetable oil

Combine all of the above in a food grinder. Now add enough nuts to form a loaf that is stiff enough to shape. Serve on a platter, with a garnish of Spanish onion rings and parsley. Flavor with desired herbs.

Enzyme Tonic Platter

1 cup tender beet or turnip
1 cup shredded cabbage
1 cup sliced carrots
1 cup diced celery

Combine all ingredients and chop or grate very fine. Then add 1 cup of any chopped nuts (peanuts are good), 1/2 cup vegetable oil. Spread out on a platter and serve promptly. Helps create a healthful enzymatic reaction.

Green Salad with Banana

3 to 4 cups lettuce leaves, cut fine
1 cup chopped nuts or seeds (or both)
2 mashed bananas

Mix together and shape into layers. Garnish with banana slices. Sprinkle with fruit juice and serve.

Cole Slaw

Shred small head of cabbage very fine. Add small amount of fruit juice, diced onion, marjoram, dill weed. Combine. Top with home-made mayonnaise dressing.

Peas 'N' Seeds

1 cup fresh peas
2 cups sliced carrots
1 small Spanish onion
1 cup sunflower seeds

Chop all ingredients very fine. Or put through a food grinder. Serve with a topping of home-made mayonnaise dressing.

Garden Soup

2 cups sweet corn, grated off the cob
1 cup assorted seeds, nuts, peanuts
4 cups sliced, peeled cucumbers
1/4 cup parsley
1/4 cup sliced celery
1/2 cup vegetable oil

Combine all ingredients in a blender. Whizz for a moment. Serve as a healthy, raw soup, brimming with good taste and enzymes.

No-Cook Enzyme Pea Soup

2 cups fresh peas
2 cups water

2 tablespoons vegetable broth powder
1 sliced tomato
1/2 avocado

Combine all ingredients in a blender. Whizz for a moment. Add vegetized salt, if desired. May be heated but only for a few moments and very lightly, too.

No-Cook Enzyme Vegetable Soup

1/4 cup washed parsley
1/4 cup chopped Spanish onion
1 cup fresh peas
1/2 teaspoon thyme
1 teaspoon marjoram
6 mashed tomatoes

Combine all ingredients and stir together until the consistency of a soup. Enjoy with whole grain bread.

High-Enzyme Swiss Breakfast

7 cups oat flakes
3 cups chopped nuts (hazlenuts are good)
1/2 cup wheat germ
2 cups sun-dried raisins
1/2 cup pitted prunes
2 grated apples
Seasonal fresh fruits

Soak this entire mixture in milk; or use fresh fruit juice, for 30 to 60 minutes. Then eat as an enzyme breakfast.

Chef's Tuna Salad

Salad greens
1 7-ounce can tuna
8 mushrooms, thinly sliced
1 hard-cooked egg, sliced
2 teaspoons drained capers
Salad dressing

Break washed, dried and crisped greens into bite-size pieces. Place in bowl. Crumble tuna into coarse pieces and scatter over greens with the remaining ingredients. Chill until serving time. Spoon salad dressing over the salad. Toss. Serve promptly.

Easy Marinated Mushroom Appetizers

1 pound medium size fresh mushrooms
3/4 cup vegetable oil
1½ teaspoons grated lemon peel
1/4 cup freshly squeezed citrus fruit juice

    1 teaspoon oregano, crushed
    1 teaspoon garlic powder

Wash mushrooms under cold, running water; drain well. Cut lengthwise through stems and buttons into 3 or 4 slices, about 1/4 inch thick; place in plastic bag or glass dish. Combine remaining ingredients; pour over mushrooms. Seal bag or cover dish; marinate for several hours in refrigerator, turning bag or stirring occasionally. Drain before serving. *Note:* Save marinade for use in a salad dressing or for marinating other foods.

Orange Plus Perfect Salad

    2 packages (3 ounces each) unflavored, sugar-free gelatin
    2 cups boiling water
    1½ cups fresh orange juice
    2 large oranges, peeled, cut into bite-size pieces (1/4 cups)
    1/4 cup yogurt
    1/4 cup chopped pecans
    Lettuce cups

Dissolve gelatin completely in boiling water; add orange juice. Chill until thick and syrupy, but not set. Drain orange pieces thoroughly; combine with about 2 1/4 cups thickened gelatin. Pour into 6 x 10 inch rectangular dish or fill 8 individual molds two-thirds full; chill. Combine remaining gelatin with yogurt; beat or blend until smooth. Fold in pecans. Carefully spoon yogurt mixture over top of orange gelatin layer; chill until firm. Cut into squares, or unmold and serve on lettuce cups.

Sun-Cooked Fruit Pie

    1 cup whole wheat flour
    1/2 cup honey
    1 cup sun-dried raisins
    6 bananas

Chop raisins very fine. Now add 1/4 cup honey and whole wheat flour. Roll out very thin. Line an oiled pie plate with this mixture. Place in the sun until dry and baked. Fill this pie shell with sliced bananas, the remaining honey and desired fruit slices.

Fish Shish Kebabs

Soak bite-size chunks of tuna and salmon in a marinade of lemon-lime juice, soy sauce, garlic, ginger, thyme, marjoram and a bay leaf. Then arrange bite-size pieces of the fish alternately with such vegetables as carrots, onions, green peppers, tomato slices, on skewers for serving.

No-Cook Fruit Bread

    2 cups whole wheat flour
    1/4 cup honey

1 cup any mashed fruit

2 eggs

Combine all ingredients in oiled bowl. Knead firmly until the consistency of good dough. Now form into a 3-inch in diameter roll. Roll this in dry whole wheat flour until it can be handled without being sticky. Wrap in wax paper. Chill overnight in refrigerator or until the bread is firm. Slice in thin slices as desired.

Cream of Grape Soup

1 pound grapes

1 cup citrus fruit juice

1/2 cup ground almonds or nuts

1 tablespoon honey

Mash the grapes. Then mix this grape mash with other ingredients. Serve chilled.

Cream of Tomato Soup

6 medium sized tomatoes

4 tablespoons ground almonds

2 tablespoons chopped parsley

Marjoram

Chives

Peel and cut tomatoes into wedge size. Now add the almonds, parsley and herbs. Add a little fruit or vegetable juice. Run through blender. Serve chilled.

Enzyme Candy Bars

1 lb. black mission figs

2 cups almonds

Wheat germ

Chop figs and almonds. Roll out on wax paper to about 1/2 inch thick. Cut into 1 inch bars. Cover with wheat germ. Wrap in wax paper and keep in refrigerator until ready to eat.

Date Pudding

2 cups whipped cream

2 cups chopped, pitted sun-dried dates

1 tablespoon lemon juice

1/2 cup chopped pecans

Combine all ingredients. Pour lemon juice over the top, or add more citrus fruit juice. Then chill until ready to eat.

No-Cook Lemon Bread

3 cups whole wheat flour

1/2 cup honey

     1 cup lemon juice
     1/2 cup yogurt or buttermilk

Combine all ingredients, except flour. When combined, add to the flour. Mix and knead until the mixture becomes a fine dough. Then roll out onto a floured board. Cut into cookie shapes. Or shape into a roll. Chill in refrigerator until firm. Then cut into slices.

Enzyme Pie

     1/2 cup sun-dried pitted dates
     1/2 cup sun-dried figs
     1/2 cup sun-dried raisins
     1 tablespoon lemon juice
     4 bananas
     2 apples, grated
     1 cup finely chopped nuts

Chop dates, figs, raisins. Now add 1 mashed banana mixed with lemon juice, Press mixture into oiled pie plate. Combine the grated apples, remaining mashed bananas and chopped nuts. Mix well together. Fill crust, with this mixture. Sprinkle top lightly with ground nuts. Makes a delicious enzyme pie that satisfies tastes while helping catalyze body cells for new slimness.

Grape Egg Nog

     1 cup cold milk
     1 egg
     1 teaspoon vanilla
     Honey, according to taste
     Black grapes

Combine milk, egg and vanilla into blender. Whizz. Then add honey and black grapes. When thoroughly blended, remove and keep chilled until ready to use.

Salmon Salad

     Slice of fresh salmon
     Honey
     Peppercorns
     Garlic
     Ginger
     Dill
     Tomatoes
     Cucumbers

Clean and de-bone salmon. Cut salmon in half. Rub with honey. Now sprinkle with mixture of peppercorn, finely chopped garlic, thinly sliced ginger and dill. Combine two pieces of fish together and place in a dish. Cover. Let remain in refrigerator overnight. When ready to serve, slice thinly and eat with sliced tomatoes and cucumbers.

Marinated Herring

    Herring
    Onions
    Vinegar
    Water
    Honey
    Mixed pickling spices

    Soak herring for 24 hours, changing water twice during the time. Cut off head, tail, scales, fins. Then slice into portion pieces. Now slice the onions. Put a layer of sliced onions and a layer of fish, layer upon layer, in an open-mouthed jar. Then combine equal amounts of water and vinegar. Add honey. Add pickling spices to taste. (Try bay leaf, chili peppers, peppercorns, cloves.) Mix up this spice combination, pour over layers of fish and onions. Let marinate some 4 hours. Then serve.

Cottage Cheese Royale

    1/2 pint skim milk cottage cheese
    1 cup chopped pecans
    1/2 cup chopped, pitted dates
    Lettuce
    Banana

    Combine cheese, pecans, dates. Thoroughly mix. Place in lettuce cups. Garnish with banana slices. If desired, drizzle with honey.

Salmon 'N' Yogurt

    Fresh salmon, sliced
    Onion rings
    Vegetized salt, to taste
    Lemon or lime juice
    Fresh dill or dill seeds
    Plain yogurt

    Combine vegetized salt, lemon juice and dill, and mix into a dressing. Pour over the salmon in a bowl. Add sliced onions. Now let marinate overnight. Before serving, fold in yogurt.

Oriental Fish-Vegetable Meal

    Halibut
    Green peppers
    Mushrooms
    Onions
    Alfalfa sprouts

Bamboo shoots
Ginger
Celery
Water chestnuts
Tuna, salmon

Use an Oriental wok (a Chinese iron cooking dish, crescent-shaped with a round bottom and two handles.) Use some oil and water in the bottom of the wok. Heat over low flame. Slice up all other ingredients. Put them into hot wok. Keep tossing as they are being heated. They will remain crisp and enzyme-potent, although tender and juicy. Serve piping hot.

Walnut "Meat" Loaf

2 carrots, shredded
1/3 cup ground oats
1/2 cup ground walnuts
2 small green onions
1 egg

Chop onions with carrots and combine with remaining ingredients. Mix well. Shape into individual servings. Enjoy with a raw vegetable salad.

Diet-O-Matic Index